Coastlines of Britain

Robert Duck

EDINBURGH
University Press

For Ann, Jennifer and Colin

© Robert Duck, 2015

Edinburgh University Press Ltd
The Tun – Holyrood Road
12 (2f) Jackson's Entry
Edinburgh EH8 8PJ
www.euppublishing.com

Typeset in 11.5/13.5 Minion Pro by
Servis Filmsetting Ltd, Stockport, Cheshire
printed and bound in Great Britain by
Martins the Printers, Berwick-upon-Tweed

A CIP record for this book is available from the British Library

ISBN 978 0 7486 9761 8 (hardback)
ISBN 978 0 7486 9762 5 (paperback)
ISBN 978 0 7486 9763 2 (webready PDF)
ISBN 978 0 7486 9764 9 (epub)

Contents

Figures

Acknowledgements

Many people have been enormously helpful to me during the writing of this work. Professor Chris Whatley and Anna Day of the University of Dundee prodded, pushed and cajoled me to do so from the outset. Professor Colin Reid of the University's School of Law helped me to locate various items of legislation within literature that were unfamiliar to me; though any mistakes in interpretation are mine alone. Professor Bob Crawford of the University of St Andrews and my colleague Dr Laura Booth's father Nigel have both kindly given me permission to reproduce photographs from their respective collections. Tracey Dixon from the School of the Environment in the University of Dundee has helped me in many ways, preparing images for publication but especially in drawing several diagrams for me. One of these (see next page) is a location map of as many of the principal places referred to in the text as her splendid cartographic skills would legibly permit. At work, I owe a huge debt of gratitude to many people, principally Jaclyn Scott, Pat Michie and Jennifer Forbes, who have tolerated me whilst I was often trying unsuccessfully to do too many things at once. Similarly, at home, where most of the writing has been done, my wife Ann has enjoyed taking virtually permanent possession of the remote control on many evenings and I want to thank her for putting up with my distracted preoccupation over the course of a year or so. All of the staff at Edinburgh University Press, in particular, Ellie Bush, John Watson, Eddie Clark and Kate Robertson, have been hugely supportive and helpful in turning my manuscript into the finished product; so too the freelance copyeditor, Jill Laidlaw. Finally, the reminiscences of a former railwayman, my Dad Bill, who began his 43-year working life with the London and North Eastern Railway Company (LNER) in 1939, continue to be a source of enormous pleasure and historical gems but, more importantly, inspiration.

Rob Duck
Tayport, April 2014

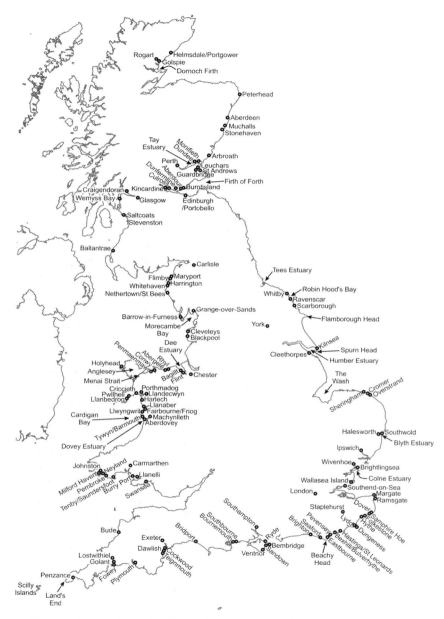

Map showing locations referred to in the text (note that, for clarity, some have been omitted) (drawing by T. Dixon).

Preface

Books don't always give the answers, you know. Sometimes they just raise the questions.
Alexander McCall Smith, *Espresso Tales*, 2005

The person that I must thank first and foremost for giving me the idea to write this book – though he will not be aware of this, nor are we acquainted – is none other than The Right Honourable Michael Portillo; but, more of that later. Since my previous book, *This Shrinking Land: Climate Change and Britain's Coasts*, was published in May 2011, a lot has happened of note around our shores. In fact, that is an understatement. Some truly dramatic events have taken place in this relatively short period and, in no particular order, my selection – and it is by no means the full menu – is as follows.

In March 2012 huge masses of chalk fell from the famous White Cliffs of Dover onto the shores of the English Channel between Dover and Deal at St Margaret's Bay in Kent.[1] The moment, in October 2011, when around 200,000 tonnes of rock fell into the sea near Hayle, close to St Ives in Cornwall, was recorded on film.[2] Dramatic 'before and after' photographs have captured the collapse of an iconic natural arch at Porthcothan Bay near Padstow and the reduction of the landmark feature to rocky rubble. Another victim of the same storms that heralded in 2014 was the sea stack of Pom Pom rock near Portland in Dorset. 'Before and after' photographs have recorded its almost instantaneous demise.[3] Nearby, a large cliff fall between Durdle Door and Lulworth Cove breached the South West Coast Path in May 2013 and sent tens of thousands of tonnes of chalk tumbling to the beach below.[4] And if that was not bad enough, a drunken man sparked a sea rescue operation after falling off the railway line alongside Poole Harbour.[5] Another rock fall in November 2011 at a leisure park at Barry in the Vale of Glamorgan left more than a dozen caravans teetering on the cliff edge.[6] Storm waves have repeatedly battered over the main railway line to south west England, where it clings to the coastal fringe at Dawlish in Devon and it succumbed, in spectacular style, to the gales of February 2014.[7] In Scotland, coastal lines between Helensburgh and Craigendoran on the Clyde coast and at Ardrossan in Ayrshire had to be suspended during the January 2014 storms, though the routes survived intact.[8] That was not the case, however, in Cumbria where the coastal railway from Carlisle to Workington was severed completely by the sea at Flimby, the embankment and ballast being washed away.[9] So too the Cambrian coast line in Wales, which was severely damaged in several places and at the time of writing (April 2014) is still out of use north of Barmouth.[10] In the first week of December 2013 low lying areas of eastern and southern England experienced their highest storm surge since the devastating tidal flooding of January 1953.[11] The slender neck of Spurn Head that projects south into the mouth of the Humber Estuary was easy prey.[12]

Numerous homes have been lost to the sea around the country including nine at Happisburgh in Norfolk which had to be demolished before they slipped onto the beach.[13] The January 2014 storms claimed a further seven cliff-top homes at nearby Hemsby, with several others being seriously undermined.[14] Further south at Easton Bavents near Sothwold in Suffolk, accelerated erosion has exacerbated the demise of other properties.[15] A retired police officer who bought her dream cliff-top house at Torquay in Devon *unseen* at auction must be having regrets. Just eight days after her telephone bid of £154,500 was accepted, a landslide in October 2012 left the 1930s property, Ridgemont House, in a precarious position, uninhabitable and worth only about £3,500.[16] In December 2012, the so-called 'Perfect Storm' hit the Scottish east coast. Strong winds along with huge waves, coinciding with high tides, caused damage from the Shetland Islands to East Lothian.[17] Stonehaven in Kincardineshire was particularly badly hit; the event was the worst in living memory, causing serious damage to waterfront properties and vehicles to be written off.[18] In the previous month, a large landslide on the North Yorkshire coast at Whitby threatened a terrace of five fisher-mans' cottages.[19] These dwellings are more than 100 years old, overlook the resort from the east side of the River Esk, and are just a few hundred metres from the historic Whitby Abbey. As a result, the properties had to be demolished. As the landslide continued to move, it began to menace more homes and also the long-closed graveyard of St Mary's Church. Human bones were exposed as the cliff gave way beneath the grounds of this ancient place of worship, said to be the oldest building in Whitby – having been founded around 1110 – and the inspiration for a scene in Bram Stoker's *Dracula*:[20]

> There was a bright full moon, with heavy black, driving clouds, which threw the whole scene into a fleeting diorama of light and shade as they sailed across. For a moment or two I could see nothing, as the shadow of a cloud obscured St. Mary's Church and all around it. Then as the cloud passed I could see the ruins of the Abbey coming into view; and as the edge of a narrow band of light as sharp as a sword-cut moved along, the church and the churchyard became gradually visible. Whatever my expectation was, it was not disappointed, for there, on our favourite seat, the silver light of the moon struck a half-reclining figure, snowy white. The coming of the cloud was too quick for me to see much, for shadow shut down on light almost immediately; but it seemed to me as though something dark stood behind the seat where the white figure shone, and bent over it. What it was, whether man or beast, I could not tell.[21]

Less well known is that Count Dracula acquired a fictional estate known as Carfax located on the very real northern bank of the outer Thames Estuary, close to the low-lying town of Purfleet:

> Dear Sirs, – Herewith please receive invoice of goods sent by Great Northern Railway. Same are to be delivered at Carfax, near Purfleet, immediately on receipt at goods station King's Cross. The house is at present empty, but enclosed please find keys, all of which are labelled.[22]

Moreover, he planned his travel there in the conventional manner of the day: 'The lamps were also lit in the study or library, and I found the Count lying on the sofa, reading, of all things in the world an English Bradshaw's Guide.'[23] Today, this once lonely spot on the Essex marshes is close to where the Channel Tunnel high-speed rail link emerges from under the Thames. This, however, is a particularly vulnerable part of Britain.[24] Exposed on the muddy foreshore at dead low water are the remains of a submerged forest of Late Neolithic age;[25] the roots from woodland of alder, ash, elm, holly and yew trees – more than 4,000 years old – are a powerful evocation of the impact of rising sea levels on 'Dracula's Garden'.[26]

In 2012, the collapse of a face cut by the sea into high sand dunes at Hemsby in Norfolk, mentioned earlier, led to two young girls being 'buried alive'. The coastguard and fire service were alerted but the girls were dug out and rescued by a retired fisherman before they arrived at the scene. This event prompted, amongst many, the following comment on the internet:

Coastal irosion [sic] it's happening alot [sic] now, and it is about time the goverment [sic] do someting [sic] about it and start protecting our coast line before it dissapears [sic] into the sea. Or have we got to wait until someone get's [sic] killed.[27]

This attracted a pithy response: 'Be good if the Government did more towards education and spelling as well!'[28]

In essence, these occurrences – all of which have happened since mid 2011 – epitomise what *This Shrinking Land* was all about; and so it goes on, they instil a sense of *déjà-vu*. The catalogue of natural 'disasters' around Britain's coast, all of which are underpinned by environmental and geological factors, is, it would seem, an endless one. So, is there room for a sequel?

I was both surprised and encouraged to receive so many complimentary letters and e-mails about *This Shrinking Land*. A recurrent theme was, why didn't I write more about, for instance, Skegness, or indeed a whole host of other places, many of which I had no first-hand experience of. People sent me examples of houses falling into the sea that I hadn't mentioned, accounts of the impacts of great storms; piers collapsing, promenades being breached by the onslaught of waves, and so on. This was hugely heartening; that folks could actually be bothered to make contact and to give me really useful material. One of the topics I had a dalliance with in *This Shrinking Land,* just one of the great ranges of impacts that railways have had on society in Britain,[29] was the enormously important role they played in bringing tourists to the coast and in the development of coastal resorts during the heyday of the Victorian era:

Railway mania in the mid to late nineteenth century saw routes constructed radiating from the industrial heartlands of Britain to numerous coastal resorts, leading to increasing numbers of both permanent residences and holiday properties.[30]

Railway posters lured would-be passengers by the curative quality of the sea air; the Great Western Railway for instance enticed visitors to Weston-super-Mare in Somerset with its promise of 'air like wine', while the Great Northern Railway famously lauded the bracing air of the Lincolnshire resort of Skegness:

For all the ills we have to bear
There's nothing cures like East Coast air
It's so bracing.[31]

The rapid growth of once sleepy, isolated villages into thriving, bustling towns – cities even – inevitably led to massive construction projects at the coast. But not only did the increasing populations need somewhere to live or to stay whilst on holiday, they needed attractions to occupy and amuse them. A sandy beach alone from which to take the air was not sufficient an allure. Thus, promenades – in effect, great sea walls at the coastal fringe, topped by walkways or esplanades – became the order of the day and no fashionable seaside town, or 'watering place', could be without one; and it had to be a long one, there was huge rivalry and competition between neighbouring resorts. Even better was a pier, extending out from the shore into the mighty sea just as far as the budget and engineering skills would allow. A public notice in *The Standard* on 21 August 1895 epitomised the lure of the coast to potential house-buyers:

Sales by Auction: Seaside Land Sale. Lee-on-the-Solent, the most rising new South Coast watering place, with extraordinary facilities of access, new railway, with station *on the very beach* close to the fine pier 750 ft in length . . . Great demand for houses all being let. Lee-on-the-Solent has a sea front view of over one mile in length, with good beach, and splendid views of the Solent and Isle of Wight; soil gravel; climate warm, dry and bracing. Luncheon at place of sale; railway tickets to be obtained of the Auctioneer at reduced fare of 5s . . .

Railways not only permitted previously unheard of human mobility, they caused our coastal settlements to grow in size at an exponential rate. However, they were also a catalyst for vast numbers of engineering projects at the coast. Structures were built with little or no regard for the environment and certainly with no thought of the impact that they might have on natural processes operating there. The imperative was to transport large numbers of passengers, thereby making money for the companies' shareholders. Railway lines themselves

Figure P.1 Volk's Electric Railway – the oldest operating electric railway in the world – laid along the back of the gravel beach at Brighton (photo: R. W. Duck).

were frequently constructed right at the coastal edge – as in the case of Brunel's London Paddington to Penzance route where it clings onto the cliff foot through Dawlish – even *directly* across beaches and mud flats. Opened in 1883, the oldest operating electric railway in the world, Magnus Volk's creation that extends for 2 kilometres along the back of the beach at Brighton, is a tourist attraction rather than a main line (Figure P.1). Though breached or seriously damaged by the sea on many occasions,[32] it has managed to survive. His later enterprise, however, the 'Daddy Long-legs' that carried passengers in an open top car between Brighton and Rottingdean, high above the waves along tracks below, survived only six days of operation before it was seriously disabled in a storm in November 1896.[33] Railways and piers are intimately associated. Indeed several piers have or have once had railways constructed along their decks. The most splendid example in Britain is that of Southend-on-Sea, at 2,158 metres in length, the longest pleasure pier in the world.[34] Overlooked from the cliff top on high by a statue of Queen Victoria benevolently pointing seawards, this projection into the Thames Estuary is one of the finest instances of engineering for the promotion of pleasure in her era. It has survived storms, being hit by

Figure P.2 The railway along the world's longest pier, at Southend-on-Sea (photo: R. W. Duck).

errant vessels and the decline in popularity of the British seaside resort but still today trains rattle back and forth carrying visitors along this mighty yet slender structure (Figure P.2). But this is a book about the physical impact of 'real' railways at the coast, as opposed to those built purely for pleasure.

The decisive idea for this sequel actually came serendipitously from watching an episode of Michael Portillo's series of television programmes of *Great British Railway Journeys*. In the closing shots of a programme filmed in Cornwall a heavy goods train, laden with china clay, was filmed passing along the western edge of the Fowey Estuary. What struck me, as the lengthy train passed slowly by the attractive village of Golant, was that the railway embankment was actually built out in the estuary, away from the natural shoreline and in so doing marooning the settlement to all but the smallest craft that could pass beneath the barrier imposed by the structure. Hereabouts the railway company had created a shorter and straighter and therefore speedier and less costly to build route by building *offshore*. It was this footage that finally made me wake up to the fact that railways have played far more important roles, both directly and indirectly, in shaping our shores that I had appreciated previously. The extreme events of the 2013–14 winter months further served to bring this stark realisation into sharp focus. This is not, however, a book about railways or railway history *per se*. Rather, it is the impacts from my geological perspective – some obvious but some far less so – that railways and railway developments have had on our island's shores that form the spinal cord of *On the Edge: Coastlines of Britain*. This is the first time that this story has been told.

Rob Duck
Tayport, April 2014

Notes

1. BBC News (14 March 2012), 'White Cliffs of Dover section falls into sea', available at http://www.bbc.co.uk/news/uk-england-kent-17366396 (last accessed 26 April 2014).

2. BBC News (11 October 2011), 'Hayle rock collapse filmed by geologist', available at http://www.bbc.co.uk/news/uk-england-devon-15250554 (last accessed 26 April 2014).

3. BBC News (7 January 2014), 'UK storms: 7 January 2014 before and after', available at http://www.bbc.co.uk/news/uk-25639777 (last accessed 26 April 2014).

4. BBC News (1 May 2013), 'Warning after massive Jurassic Coast cliff fall', available at http://www.bbc.co.uk/news/uk-england-dorset-22371787 (last accessed 26 April 2014).

5. BBC News Dorset (4 October 2011), 'Drunk man falls off railway line into sea off Dorset', available at http://www.bbc.co.uk/news/uk-england-dorset-15166990 (last accessed 26 April 2014).

6. BBC News (2 November 2011), 'Landslip leaves 15 Barry caravans teetering over cliff', available at http://www.bbc.co.uk/news/uk-wales-south-east-wales-15540048 (last accessed 26 April 2014).

7. BBC News (5 February 2014), 'Devon and Cornwall storm causes "devastation"', available at http://www.bbc.co.uk/news/uk-england-26044323 (last accessed 26 April 2014).

8. *The Scotsman* (20 January 2014), 'Weather: Wettest month ever as storms hit again', available at http://www.scotsman.com/news/environment/weather-wettest-month-ever-as-storms-hit-again-1-3255615 (last accessed 26 April 2014).

9. BBC News Cumbria (6 January 2014), 'Repairs to storm-hit Cumbrian rail line to "take a week"', available at http://www.bbc.co.uk/news/uk-england-cumbria-25612478 (last accessed 26 April 2014).

10. BBC News Wales (4 February 2014), 'Cambrian coast rail flood repairs at Barmouth and Pwllheli to take months', available at http://www.bbc.co.uk/news/uk-wales-26041688 (last accessed 26 April 2014).

11. Met Office and Centre for Ecology & Hydrology (February 2014), *The Recent Storms and Floods in the UK*, available at http://www.metoffice.gov.uk/media/pdf/n/i/Recent_Storms_Briefing_Final_07023.pdf (last accessed 26 April 2014).

BBC News (6 December 2013), 'Lethal storm and tidal surge sees thousands out of homes', available at http://www.bbc.co.uk/news/uk-25220224 (last accessed 26 April 2014).

12. BBC News Humberside (10 December 2013), 'Public warned off Spurn Point after tidal surge damage', available at http://www.bbc.co.uk/news/uk-england-humber-25317870 (last accessed 26 April 2014).

13. BBC News (11 April 2012), 'Cliff top homes demolished in Happisburgh', available at http://www.bbc.co.uk/news/uk-17685181 (last accessed 26 April 2014).

14. BBC News (6 December 2013), 'Norfolk floods: Seven Hemsby homes badly damaged by waves', available at http://www.bbc.co.uk/news/uk-england-norfolk-25254808 (last accessed 26 April 2014).

15. BBC News Suffolk (1 January 2014), 'Easton Bavents cliff top house facing demolition up for sale' available at http://www.bbc.co.uk/news/uk-england-suffolk-25375207 (last accessed 26 April 2014).

16. Mail Online (23 October 2012). 'I knew I should have ordered a surveyor's report', available at http://www.dailymail.co.uk/news/article-2221965/Torquay-house-left-teetering-cliff-edge-landslide-week-disabled-Sue-Diamond-paid-154k-it.html?ICO=most_read_module (last accessed 26 April 2014).

17. BBC News Scotland (16 December 2012), '"Perfect storm" hits Scottish coast', available at http://www.bbc.co.uk/news/uk-scotland-20750940 (last accessed 26 April 2014).

18. STV News (2012), 'Stonehaven storm 2012', available at http://news.stv.tv/galleries/stonehaven-storm-2012/ (last accessed 26 April 2014).

19. BBC News (28 November 2012), 'Whitby landslip: Five houses left dangling over drop', available at http://www.bbc.co.uk/news/uk-england-york-north-yorkshire-20527471 (last accessed 26 April 2014).

20. BBC News (10 January 2013), 'Whitby landslip exposes human bones at "Dracula graveyard"', available at http://www.bbc.co.uk/news/uk-england-york-north-yorkshire-20970716 (last accessed 26 April 2014).

21. Stoker, Bram (1897), *Dracula*, Edinburgh: Archibald Constable and Company.

22. Stoker, Bram (1897), *Dracula*, Edinburgh: Archibald Constable and Company.

23. Stoker, Bram (1897), *Dracula*, Edinburgh: Archibald Constable and Company.

24. Duck, R. W. (2011), *This Shrinking Land: Climate Change and Britain's Coasts*, Dundee: Dundee University Press.

25. Murphy, P. and Trow, S. (2005), 'Coastal change and the historic environment', *Conservation Bulletin*, 48, 8–12, English Heritage.
    Murphy, P. (2009), *The English Coast: a History and a Prospect*, London: Continuum International Publishing Group, 282 pp.

26. McFadyen, Jock (2001), *Purfleet from Dracula's Garden,* Tate, oil on canvas.

27. Kendrick, K. (2012), 'Schoolgirls survive being buried alive after beach cliff collapse', available at http://www.parentdish.co.uk/2012/04/11/schoolgirls-survive-being-buried-alive-after-beach-cliff-collapse/ (last accessed 26 April 2014).

28. Kendrick, K. (2012), 'Schoolgirls survive being buried alive after beach cliff collapse', available at http://www.parentdish.co.uk/2012/04/11/schoolgirls-survive-being-buried-alive-after-beach-cliff-collapse/ (last accessed 26 April 2014).

29. Evans, A. K. B. and Gough, J. V. (2003), *The Impact of the Railway on Society in Britain: Essays in Honour of Jack Simmons*, Aldershot: Ashgate, 315 pp.

30. Duck, R. W. (2011), *This Shrinking Land: Climate Change and Britain's Coasts*, Dundee: Dundee University Press.
31. Great Northern Railway Poster (1908), *Skegness 'It's so Bracing'*.
32. See, for instance: 'The gale', *The Times*, 12 September 1903.
33. Volk's Electric Railway Association, available at http://volkselectricrailway.co.uk/ (last accessed 26 April 2014).
34. Natonal Piers Society: History of Southend-on-Sea Pier, available at http://www.piers.org.uk/pierpages/NPSsouthend.html (last accessed 26 April 2014).

1

Cometh the Railway

The first shock of a great earthquake had, just at that period, rent the whole neighbourhood to its centre. Traces of its course were visible on every side. Houses were knocked down; streets broken through and stopped; deep pits and trenches dug in the ground; enormous heaps of earth and clay thrown up; buildings that were undermined and shaking, propped by great beams of wood. Here, a chaos of carts, overthrown and jumbled together, lay topsy-turvy at the bottom of a steep unnatural hill; there, confused treasures of iron soaked and rusted in something that had accidentally become a pond. Everywhere were bridges that led nowhere; thoroughfares that were wholly impassable; Babel towers of chimneys, wanting half their height; temporary wooden houses and enclosures, in the most unlikely situations; carcases of ragged tenements, and fragments of unfinished walls and arches, and piles of scaffolding, and wildernesses of bricks, and giant forms of cranes, and tripods straddling above nothing. There were a hundred thousand shapes and substances of incompleteness, wildly mingled out of their places, upside down, burrowing in the earth, aspiring in the air, mouldering in the water, and unintelligible as any dream. Hot springs and fiery eruptions, the usual attendants upon earthquakes, lent their contributions of confusion to the scene. Boiling water hissed and heaved within dilapidated walls; whence, also, the glare and roar of flames came issuing forth; and mounds of ashes blocked up rights of way, and wholly changed the law and custom of the neighbourhood.

In short, the yet unfinished and unopened Railroad was in progress; and, from the very core of all this dire disorder, trailed smoothly away, upon its mighty course of civilisation and improvement.

Charles Dickens, *Dombey and Son*, 1848

Mania

When Charles Dickens wrote *Dombey and Son* Britain was in the throes of railway mania. In the mid to late 1840s lines were being proposed and constructed with feverish pace all over the country, into even the remotest regions. In the north London suburb of Camden Town, where he was resident at the time, Dickens likened the coming of the railway to a natural disaster, – an earthquake – was this the destructive price that had to be paid for improvement? Little did Dickens realise that some 15 years later in 1863 he was actually to experience a real earth tremor, large by British standards, in the south of England. Furthermore, two years later on 9 June 1865, he was to survive a dreadful accident at Staplehurst on the South-Eastern Railway in Kent.[1] Bound for London from Folkestone, while returning from France, the first seven carriages of the train Dickens and his companions were travelling in became derailed as it jumped a gap in rails that were being repaired at the time and plunged off the side of a cast iron viaduct. The only first class carriage to remain on the track was the one in which he was travelling. The accident, sadly not unusual in its day, claimed ten lives and a further forty people were injured, many of them seriously.[2] In a letter to his doctor the following day, Dickens noted: 'I was in the terrible accident yesterday, and worked some hours among the dying and dead. I was in the carriage that did not go down, but hung in the air over the side of the broken bridge, I was not touched – scarcely shaken.' He signed off: 'Ever Yours C.D. (I can't sign my flourish today!)'.[3] Before leaving the scene, he allegedly remembered that he had left his unfinished manuscript of *Our Mutual Friend* on the train and he returned to the dangerously unstable carriage to retrieve it.[4] Exactly five years later to the day, Dickens passed away.

In London, the underground system was also being developed in the 1860s, resulting in the digging of huge trenches by those companies, like the Metropolitan Railway, that employed the 'cut and cover' method. As a result, countless streets became vast trenches, were boarded off for years and there was widespread devastation in the capital.[5] In Edinburgh the situation was similar and by the close of the nineteenth century railway lines had proliferated throughout the city prompting the correspondent, *Cockburnite*, to note in 1894 with sarcastic disdain that:

> . . . Edinburgh may now be considered to be merely a railway station, or siding, on the road to the north. Has the city surrendered its

historical position of a capital, and its municipal integrity as a town to a modern railway company? Future gazetteers may have to insert in their list – Edinburgh, a provincial railway station, with a castle and a bridge, on the North British Railway.[6]

Indeed, the country-wide ruthlessness with which the railway companies drove cuttings through or built huge viaducts above our cities destroying all in their wake for the sake of progress, is well known. But what impact did the railways have at the periphery of Britain, along our coasts, away from the major urban centres? Many people did not even want to find this out and went to considerable lengths – usually in vain – to prevent the railway's arrival. The architect and engineer, George Knowles, who lived in a splendid house on Scarborough's South Bay, published a pamphlet in 1841, at his own expense, detailing how the watering place's respectability would be ruined by an influx of the wrong sort of people:

The inhabitants of the place are well pleased to see respectable people come amongst them; but they have no wish for a greater influx of vagrants; and those who have no money to spend; and I am sure that our respectable visitors have no relish for either a railroad, or the pleasure of such company; on the contrary they generally express their disapprobation of the measure; and I have heard very many of them say, that if there was a railroad to Scarborough, they should never come again, as visitors on pleasure.[7]

Knowles saw no advantage whatsoever in a reduced journey time from York; 'the saving of two hours' time would be all they could expect from it'. Cutting the journey time by around two thirds, 'I cannot think . . . would be any inducement, or would have any tendency to increase the number of those who only come hither to breathe the pure air, and drink the waters of Scarborough'. Moreover, '. . . in a few years more, *the novelty of not having a Railroad*, will be its greatest recommendation.'[8] In desperation, Knowles published a supplement to his pamphlet three years later warning the inhabitants of the town that they should be prepared for;

. . . an enlargement of the jail, for confining a greater number of thieves, pick-pockets, vagrants, which a railway would most certainly entail upon them; – then this 'Queen' of watering places,

would have to weep over her fallen greatness, and rue the day that
she ever saw her shores invaded by a railroad.[9]

The line to Scarborough from York, engineered by George Hudson,
the so-called Railway King, opened a year later in 1845.

Whilst the route to Scarborough from the interior heartland of
Yorkshire, in common with so many others, opened up access to and
from the resort, it did not impact upon the natural configuration of
the coastline. That was not the case elsewhere, however. The so-called
'reclamation' of coastal lowlands by means of dykes or embankments,
to become the foundations for straight sections or gently curving railways, thereby smoothing out the intricacies and complexities of our
coastal contours, was clearly an attractive option to nineteenth-century
engineers. It would enable routes to traverse mainly low-lying lands
without harsh gradients and, moreover, would link together coastal
communities and watering places. In addition, coastal routes would
often avoid expensive cuttings or tunnelling, though this was by no
means always the case as protruding rocky headlands could provide
obstacles. The very term, reclamation, implies – quite incorrectly – that
humans were taking back land that had previously been 'stolen' by
the sea and it epitomises the engineering attitude of the time. Today,
the far more appropriate term land 'claim' should be adopted but the
concept of land reclamation has become ingrained. Some seven years
or so before the epidemic of railway fever really took hold in Britain,
John Rooke,[10] the self-taught writer on political economy and geology,
made an early proclamation:

> And as arms of the sea and estuaries often cut off the best and
> most direct lines of railroads, so the most ready and skilful mode
> of reclaiming these from the sea, embrace a leading branch of our
> subject. Viewed in this light, Geology becomes of momentous
> importance, as a basis on which these considerations in some
> measure depend.[11]

Moreover, Rooke was to be proved right. His interest in the practical applications of geology had become aroused by the prospect
of a railway across Morecambe Bay in Lancashire to link with west
Cumberland when he met the line's promoter, the engineer and
linguist Hyde Clarke, in 1836. Coastal lands were popular to railway
developers not only because of their low altitude and often relatively

flat topography; they could often be acquired far more readily than sites at some distance from the sea. It was to the railway developers' advantage that the boundary between land and sea has always been rather a 'grey area' in legal terms. This is because coasts can be highly dynamic, mobile and can change dramatically over short time scales, whereas the lines that are drawn on maps to denote land ownership are fixed. For instance, sediment may build up in one area of coast so that inter-tidal mud or sand flats become more extensive both laterally and vertically, eventually becoming salt marshes supporting luxuriant vegetation. By contrast, another area, where the geology is of 'soft' materials, may be characterised by vigorous erosion and rapid cliff recession. Where there is natural coastal accretion, a landowner will apparently gain property, whereas natural erosion has the opposite effect. Nonetheless, the lines drawn on maps to reflect such property boundaries remain unchanged until such time as the area in question is re-surveyed; hence the 'grey area' in terms of who actually owns what.[12]

The term 'foreshore' is often used synonymously with inter-tidal zone or beach, though this is not quite accurate and, indeed, the precise definition of the foreshore differs slightly according to English and Scots law.[13] What is important is that the owner of the foreshore – however it is precisely defined – is, in today's simple terms, the Crown. In some parts of Britain the foreshore will be gaining in extent by accretion, in others it will be decreasing in extent by erosion.[14] Furthermore, in mid-nineteenth century Britain, the Crown quite simply did not value the foreshore in today's property terms. It was not considered as an asset; rather it was considered as waste land of little or even no monetary value. Moreover, as its precise extent could not be ascertained at any particular site, the Crown did not really know exactly which part of the coastal zone it owned. As such, the foreshore was very attractive to aggressive railway companies – it could be acquired with very little difficulty. The Crown did not value it and was minded to dispose of it even if it did not in fact, as a consequence of coastal dynamism and mobility, necessarily own all of the property it was handing over to the railway companies. The Crown was single-mindedly interested in *progress* during the Industrial Revolution; beaches, inter-tidal flats, salt marshes and slob lands were regarded simply as waste lands (see Chapter 4) that needed to be put to productive use and thus in many places foreshores became the foundations of Britain's railways on the edge. *The Railways Clauses Consolidation*

Act 1845,[15] along with its equivalent legislation for Scotland,[16] super-seded what were originally private Acts of Parliament to create new railways. The focus of Section 17 of this Act was on: 'Works below High-water Mark not to be executed without the consent of the Lords of the Admiralty'[17] and, specifically for Scotland, 'Works on the Shore of the Sea, &c. not to be constructed without the Authority of the Commissioners of Woods and Forests and Commissioners of the Admiralty'.[18] The full details are as follows:

It shall not be lawful for the Company to construct on the Shore of the Sea, or of any Creek, Bay, Arm of the Sea, or navigable River communicating therewith, where and so far up the same as the Tide flows and re-flows, any Work, or to construct any Railway or Bridge across any Creek, Bay, Arm of the Sea, or navigable River, where and so far up the same as the Tide flows and reflows, without the previous Consent of Her Majesty, Her Heirs and Successors, to be signified in Writing under the Hands of Two of the Commissioners of Her Majesty's Woods, Forests, Land Revenues, Works, and Buildings, and of the Lord High Admiral of the United Kingdom of Great Britain and Ireland, or the Commissioners for executing the Office of Lord High Admiral aforesaid for the Time being, to be sig-nified in Writing under the Hand of the Secretary of the Admiralty, and then only according to such Plan and under such Restrictions and Regulations as the said Commissioners of Her Majesty's Woods, Forests, Land Revenues, Works, and Buildings, and the said Lord High Admiral, or the said Commissioners, may approve of, such Approval being signified as last aforesaid; and where any such Work, Railway, or Bridge shall have been constructed it shall not be lawful for the Company at any Time to alter or extend the same without obtaining, previously to making any such Alteration or Extension, the like Consents or Approvals; and if any such Work, Railway, or Bridge shall be commenced or completed con-trary to the Provisions of this Act, it shall be lawful for the said Commissioners of Her Majesty's Woods, Forests, Land Revenues, Works, and Buildings, or the said Lord High Admiral, or the said Commissioners for executing the Office of Lord High Admiral, to abate and remove the same, and to restore the Site thereof to its former Condition at the Cost and Charge of the Company; and the Amount thereof may be recovered in the same Manner Penalty is recoverable against the Company.[19]

In simple terms, it was the case that railway companies had only to obtain explicit permission from the Crown to build across the foreshore. If they did not do so, any structures would be removed by the Crown and the environment restored to its former state at a cost charged to the railway company concerned. This was therefore a piece of legislation that was entirely in tune with the aims and aspirations of nineteenth century railway builders. Indeed it played straight into their ruthless hands.

Away from the coast, the purchase of land was often far more problematic. The excuses put forward to account for the expense of land acquisition in Scotland were summed up beautifully in 1848 by a correspondent to *Herapath's Railway and Commercial Journal*:

The greatest difficulty railway interests have to contend with in Scotland is the unreasonable demands, – nay, the avaricious cupidity of the owners of the soil. The most sterile patch of the Ochills [sic], the bleakest moor in Nithsdale, the craggiest acre within the shade of the Grampians, touch it by a railway, and you will hear of land valued at £3 per acre that never yielded as much grass as would suffice for a week's summer keep to a Shetland pony. Spoiling the view from the hall, 'laying open the privacy' (a favourite phrase) of a gentleman's demesne, coming too close to a rabbit warren, or smoking the 'gude wife's' washing, are but a few of the items manufactured in 'railway claims shops'.[20]

Little wonder that railway companies both north and south of the border often chose to keep the cost of land as low as possible, even though construction near the coastal edge brought a wide variety of engineering challenges with equally widespread and vociferous public opposition.

Hyde Clarke's Morecambe Bay scheme was to come to fruition in 1857, thanks to the engineering skills of James, later Sir James, Brunlees. He had gained valuable prior experience in Northern Ireland where he had constructed a railway embankment, part of the Londonderry and Coleraine Railway, across Rosse's Bay on the southern side of the Foyle Estuary. This structure was built 'with considerable difficulty owing to the unusual depth and treacherous nature of the alluvial deposit around the bay.'[21] Crossing inter-tidal lands, salt marsh, sandflat and mudflat, skirting the northern rim of Morecambe Bay on embankments and with viaducts spanning the large estuaries of the influent Rivers Leven and

Kent, the Ulverston and Lancaster Railway was an even more ambitious project.[22] However, it was by no means a one-off in Britain.

Where are the sands at Grange-over-Sands?

Morecambe Bay is an arm of the Irish Sea, situated to the west of Lancaster; its average width is about twelve miles and it extends about seventeen miles inland from the mouth. Its dangers are well known to all who travel the over-sands route, on account of the constant shifting of the fresh-water channels, and the treacherous nature of the sands, more especially during freshes in the rivers.

Sir James Brunlees, *On the Construction of the Sea Embankments across the Estuaries Kent and Leven, in Morecambe Bay, for the Ulverstone* [sic] *and Lancaster Railway*, 1855

Morecambe Bay has become infamous in recent years owing to the tragic deaths of at least twenty-one Chinese cockle pickers in February 2004 who were drowned by the incoming tide.[23] In large shallow embayments such as this, the advancing tide can move with great rapidity, catching people unawares; it is an area of the British coast that has claimed many lives over the centuries but never on such a large and terrifying scale. Located at the mouth of the Kent Estuary, one of several rivers that discharge into the bay, the fashionable and refined watering-place of Grange-over-Sands developed on the northern shore during the Edwardian era. Its very name evokes the image of a splendid golden beach. However, nothing could be further from the case. The resort was originally situated directly on the shoreline but the construction of an embankment carrying the Ulverston and Lancaster Railway,[24] which opened to traffic in 1857, effectively divorced it from the sea, leaving it to develop on the landward side of the barrier. This was not a rapid process and, as in other localities, stagnant pools became isolated from the sea by the railway. Some six years after the opening of the line, a call was being made in the local press for the filling up and levelling of the very unsightly hollow between the railway embankment and Lower Grange Road along with other improvements to the town to bring it up to the standard of watering places such as Teignmouth and Dawlish in Devon.[25]

An editorial in the *British Medical* Journal in 1873, was hugely critical of the place:

The inhabitants of the Lancashire watering-place seem, according to the information which reaches us, to be pursuing just now a very suicidal policy in regard to their future interests. Grange is situated on the coast line of the Furness Railway around Morecambe Bay, but cut off by the railway from the shore. Consequently it has no bathing, no sands, and lacks many inducements for strangers.[26]

It was perhaps rather unkind to level this criticism at the inhabitants of Grange themselves who had had little or no say in the fate of their village once the eventual route of the railway had been chosen. The 'very unsightly' hollow had subsequently been landscaped: 'Inside the rail, a portion of the pre-existing shore has been converted into a kind of ornamental water, below high-water mark.'[27] Today, few people who sit or walk around this secluded spot – with a fountain at its heart but no vista of the sea owing to the railway embankment – will likely realise that it was once a part of the wave-lapped shore (Figure 1.1). The railway embankment was also a barrier to free drainage and this was certainly

Figure 1.1 Ornamental garden at Grange-over-Sands, once part of the beach but now isolated from the sea by the railway embankment. The original shoreline is behind the trees to the right (photo: R. W. Duck).

not conducive to Grange's growing reputation as a resort for invalids with bronchial and pulmonary complaints owing to the mildness of its winter climate. Pulling no punches, the *British Medical Journal* remarked that:

> The only drawback, for invalids a very great one, is its want of sanitary arrangements. It has no drains . . . and water closets running into pervious cesspools, or, what is worse, into chinks in the rocks, which are of limestone, and broken into wide, cavernous openings.[28]

Thus sewage, no longer being discharged into the sea owing to the railway, became concentrated in the pools to landward of the embankment. Unfortunately, the chief spring and water supply to the village emanated from the Carboniferous limestone rocks within what had become the ornamental gardens: 'With such arrangements of cesspools, &c., it is impossible to suppose such [a] spring can long remain untainted by the percolation of drains and water closets on the hill side above; and, indeed it is already in a very dubious condition.'[29] The latter point was, needless to say, disputed strongly by local residents.[30] Railways have certainly had unforeseen effects at the coast.

As in many places around Britain, another concern was that the railway formation impeded access to the shore. Nearly 20 years after its opening, local property owners and ratepayers were still campaigning for bridges to be built over the line. In 1876 they penned an open letter to his Grace the Duke of Devonshire and the Directors of the Furness Railway Company (the 7th Duke was the major investor in the Company, which had bought out the Ulverston and Lancaster Railway in 1862):

> We, the undersigned, owners of property, residents, and others interested in the prosperity of Grange-over-Sands, beg leave to represent the great importance of safe and easy access to the shore. The level crossings, by which alone it can at present be reached, are inconvenient and dangerous.[31]

The letter went on to recommend that two bridges should be provided, to the east and the west of the village, 'so as to afford free access at all times for the enjoyment of fresh air brought in by the tide, with the facilities which would probably follow for bathing and boating'. The residents trusted that the Duke and the Directors would listen to their earnest request that:

the impediment which we have pointed out to the full enjoyment and more general attraction of a watering place so healthy and agreeable, may be removed, and a prospect opened of amusement, the want of which has been seriously felt and complained of.[32]

In the late nineteenth century, Grange's advantage as a resort, whilst cut off from the sea, was attributed to the presence of the Kent Estuary:

which secures a fine tide for the benefit of visitors at all times of the year, and also enables Grange and Morecambe to keep up daily communication by means of boats in summer. The railway effectually protects the lower part of Grange from the inroads of floods in autumn and winter, but at the same time rather stands in the way of communication with the shore . . . [33]

A plaque on the promenade wall, built to seaward of the embankment by the Furness Railway Company in 1902, commemorates the site of Bayley Lane Pier built by the Morecambe Bay Steamboat Company in 1870. It notes that remains of this structure can be seen jutting out of the sands but accretion of salt marsh has rendered these no longer obvious. Nearby was Clare House Pier, built in 1893 by a local businessman; by 1910 it had fallen into disuse and was blown down by a gale in October 1928.

Construction of the sea embankments – which mainly utilised sand sourced locally from the shores of Morecambe Bay, as was the normal practice of the day – proved to be more difficult and took longer than anticipated.[34] Problematical ground conditions were often compounded by the coincidence of violent gales with high spring tides. However, eventually some very large areas of land claim were associated with the construction of this coastal railway including the loss of extensive swathes of salt marsh;[35] indeed the Ulverston and Lancaster were able to recoup some of their losses through the sale of land claimed from the bay.[36]

Nothing particular strikes the eye of the traveller until the embankment across the Milnthorpe Sands is reached, which, to the inexperienced in such matters, will no doubt create some degree of pleasurable surprise. The embankment is formed by sea sand, which, during the process of making was protected by a guard of stones. It was then covered with a coating of clay, and afterwards by a bed of broken stones, and finally the facing was pitched with blocks of stones, forming a most firm and finished causeway, with a slope capable of resisting the roughest and heaviest tides which occasionally rake its surface.[37]

Despite its alleged durability, the railway embankment itself succumbed to one of the 'roughest and heaviest tides' in 1918, necessitating a major repair:

Serious damage was caused to the Furness Railway line near Grange-over-Sands early yesterday morning. During a high tide in Morecambe Bay the sea washed away the embankment for 60ft., undermined the railway, and left a huge gap, in some places 45ft. deep. The water surged over the rails and through the gap on to the roadway, which was flooded. Happily the damage was discovered in time and traffic was suspended.[38]

Similarly, in the unusually wet weather of October 1954, associated with gales and high tides, holes 20 feet wide and 10 feet deep were torn out of the stone pitching that protected the line.[39]

Land claim and the building of the huge embankment were to reduce estuarine capacity and to have a profound effect on the current patterns in the bay; in particular a change in the orientation of the Kent's main channel and a reduction in the velocity of tidal currents on the ebb tide led to the deposition of substantial quantities of mud.[40] Once the mud began to accumulate at Grange-over-Sands it began to be colonised by vegetation, principally marsh cord grass (*Spartina* spp.), and eventually transformed into luxuriant salt marsh (Figure 1.2). This is now used to graze sheep and sometimes cattle. Inundated on only the highest of spring tides, this green mantle is criss-crossed by sheep tracks leading to several drainage creeks that have cut down through the deposits of mud. The once sandy beach that gave the resort its name has now gone for ever and it is widely asserted that the most significant factor to affect this part of Morecambe Bay was the construction of the railway.[41] The associated claim of around 1,000 acres of land has changed the outline of the bay more than any other recent event. Salt marsh accretion continues to this day and in terms of managing the coast, a salt marsh helps to provide a natural defence against wave attack. However, a wave-washed sandy beach would have provided Grange-over-Sands with a much higher level of tourist amenity.

Figure 1.2 The expanse of salt marsh at Grange-over-Sands that has accumulated to seaward of the sea wall and railway line (photo: R. W. Duck).

Lines on the edge

The Fowey Estuary in Cornwall has a long history of industrial activity including nineteenth century tin mining in its catchment area and occupies a beautiful steep-sided and wooded valley.[42] The water body is a splendid example of a ria. As such, it is a coastal inlet formed by the partial submergence of a river valley and, like many such features, it is tree-like in plan with several branches or embayments along either bank. In this part of the world these small tidal bays are known as pills. The railway proposed was to extend to the deep water port of Fowey, branching off from the main Great Western route at the town of Lostwithiel: 'A great portion of the intended route of this projected railway lies on the banks of the Fowey river, which are seldom visited by tourists and others . . . '[43] Indeed the route chosen was right along the western shoreline of the estuary with the pills being traversed by means of embankments so as to keep it as straight as possible. At the small village of Golant, roughly midway between Lostwithiel and Fowey, the single-track line was to cross Golant Pill on a narrow embankment:

> Messrs Mead and Lang, of Liskeard, are the contractors; their cashier Mr Cook obligingly afforded information. The 'Golanters' as the villagers are termed, are somewhat annoyed at finding that the pretty view of the river from their village will be partially hid by an embankment passing over the sand by its side for some distance. Openings, however, must be left for the passage of boats into the river from the shore.[44]

At the time of completion of the line in 1868, prior to its opening one year later, Golant was described as a village 'too small to have any claims to a station'.[45] The line carried goods for the first 10 years of its life, after which it fell into a period of decline and dormant disuse. However, in 1895 it was reinvigorated and opened for 'passengers, goods and merchandise traffic'; moreover, Golant acquired its long-awaited halt.[46] This small rural community, it would seem, expressed little further dissent about the embankment and indeed the intrusion of the structure provided the village with a safe harbour, albeit for craft of a rather restricted size (Figure 1.3). Today, lengthy trains laden with china clay are the only traffic to still rumble over this slender, man-made isthmus. Indeed it has been wryly noted that the price Golant has had to pay for its sheltered haven is a lack of public transport links.[47] Larger boats are obliged to moor outside the railway formation. A kilometre to the south, the embankment has similarly isolated the remote Bodmin Pill. Immortalised in *The Ballad of Bodmin Pill* by the British rock band New Model Army, this inlet drains through an arch in the railway at low tide to expose a broad mudflat though which a muddy creek meanders. A former seventeenth-century watermill on the landward side of the railway now rather fittingly houses a sound recording studio. Here, at Golant and elsewhere, the railway embankment has served to modify the natural shoreline configuration of the western margins of the estuary and in so doing has modified the natural patterns of currents and sedimentation.

Further to the north-east on the trunk line of the Great Western system in Devon a similar embankment has created what is today called Cockwood Harbour. The line between Exeter and Dawlish Warren via Starcross hugs the western shores of the estuary of the River Exe. At Cockwood, marshy ground, known in the nineteenth century

500 m   N

Figure 1.3 Aerial image of the Fowey Estuary and Golant. The railway embankment along the western edge of the ria has cut off Golant Pill where it crosses the mudflats. Further south, Bodmin Pill has also been completely cut off by the railway (© 2014 Google – DigitalGlobe, Getmapping plc, Infoterra Ltd & Bluesky – Imagery January 2005).

as Cockwood Swamp, had been largely reclaimed for agriculture in 1808 but its tidal mouth at the east remained.[48] To avoid an inland detour, the line traverses this inlet on a large embankment thereby creating the harbour to the west. Today a narrow gap under a bridge in the embankment provides the only connection of the harbour with the Exe for boats. This was a difficult structure to build;[49] a challenge for Isambard Kingdom Brunel, whose feats of Great Western Railway engineering, and their impacts, further south along the coast feature later in the book. The original plan to build an embankment across this inter-tidal zone was a failure as firm foundations could not be found, even by excavating to a depth of 40 feet.[50] Eventually, and with much difficulty owing to severe gales, high tides and piling though thick sequences of soft mud, a wooden viaduct was built over the swamp, which was completed in December 1845.[51] Vulnerable to subsidence and to south-easterly gales,[52] eventually this had to be replaced with the more solid, yet much-repaired, stone-faced embankment seen today, a feature that has dramatically altered the appearance of Cockwood and the western shoreline of the Exe (Figure 1.4). Elsewhere in Britain, coastal railway schemes were to have much greater impact and received much more vociferous opposition, though invariably they went ahead despite protestations and objections, such was the steely determination of the railway companies.

Figure 1.4 The railway embankment in the Exe Estuary isolating Cockwood Harbour from the western shore. The view is east across the estuary towards Exmouth (photo: R. W. Duck).

The Firth of Forth of eastern Scotland has a long history of land claim around its fringes. Extensive areas of inter-tidal mudflat and salt marsh areas had been claimed from the sea for agriculture by 1840, typically by the construction of earth embankments (dykes) with cores of impermeable clay. These were faced on the water side with cut stone blocks to prevent wave erosion. Subsequent land claim schemes in the Forth that continued into the 1920s were principally for industrial rather than agricultural purposes, including the provision of land for railways. These typically used nearby colliery and oil shale mine waste to raise the land level behind the retaining embankments.[53] In total, since the beginning of the seventeenth century, it is estimated that about half of what was naturally inter-tidal land in the Forth has been lost by human intervention.[54] The railway arrived comparatively late at the former Royal Burgh of Culross on the northern shore (Figure 1.5). Famous for its sixteenth-century Palace and seventeenth-century Town House, this village, historically a centre of coal mining and salt pans, has been described as 'the nearest thing to a sixteenth-century time capsule anywhere in Scotland.'[55] Whilst the Firth of Forth was the 'chief theatre' of

Figure 1.5 The railway embankment at Culross – much repaired – cuts off the town from the northern shore of the Firth of Forth (photo: R. W. Duck).

Scottish salt production, Culross, in particular, was perhaps the leading salt manufacturing vicinity in the latter part of the sixteenth century. However, it never recovered from a great storm in 1625 which ravaged the Forth.[56] The decline of industry along the Culross shore is reflected in the guide *The Fringes of Fife*, wherein the journalist John Geddie described this spot in 1894 as:

> . . . a nook of Fife difficult to get at, and still harder to get away from. Railways do not come within two miles of it; no passengers by water land at its little pier. The wooded ridge sweeps round it so steeply and closely that the trees and hanging gardens look directly down upon the Sandhaven and the West Green.[57]

Further, he reflected on the area's industrial history that: 'The seams are worked out, and the coal-heughs [shafts] closed. Of Culross's fifty saltpans not one sends up its smoke. They are things of the past . . . '[58]

However, by the time Geddie updated and enlarged his work in 1927, the railway that hugs the northern shore of this great firth, along much of its route between Kincardine and Dunfermline, had become operational in 1906. As a result, Geddie was obliged to revise his idyllic perception of post-industrial Culross and the neighbouring village of Low Valleyfield to the east, noting with considerable regret that:

> The railway has drawn its score between the old causeys [causeways] and biggings [buildings] of Culross and the sea; and it has similarly treated Low Valleyfield and its adjuncts – a convenience doubtless, but, at many points, a sad break in historical continuity and in pictorial harmony of effect.[59]

This was a reference to the fact that, once again, a railway line had been built on an embankment along the foreshore, cutting off the communities – both physically and visually – from the very water body on which their livelihoods had depended in so many ways for centuries. Remains of the once extensive Culross saltworks and coal mining activities were also all but obliterated by the railway; so too was the old pier, ruined as it was driven through by the embankment.[60] Mindful perhaps of the 'Sandhaven', Geddie mused wistfully that: 'Good times would return, and the summer bather and tripper would flock to Culross Bay, could only some kind magician exchange its mud for the clean white sands of yore.'[61]

Local residents had been outraged at the prospect of the railway embankment slicing across the Culross foreshore. Despite their protests on a variety of counts, they could not stop it happening: 'But though it [the embankment] is to be a dwarf, and not a giant, it will still effectually spoil the beauty of a lovely coast road, and grievously interfere with the rights of the general public.'[62]

The reference to the height of the structure was in comparison with the huge embankment built some four decades earlier along the shore at Granton near Edinburgh. This feature was so high, ugly and obtrusive that it completely obliterated the view of the sea from nearby houses and was the scene of a disaster in 1860 that features in Chapter 2:

Now what is proposed to be done [at Culross] is to run a railway embankment to carry the proposed line along this lovely marine promenade and between it and the sea. If any one wishes to see the full effect of such an abomination, let him look at the roadway between Trinity and Granton, where one of the most beautiful seaside roads in the country has been utterly and irretrievably ruined.[63]

Loss of the pier, the beach and the overall amenity at Culross were real concerns and a source of frustration:

Starting from Torry Point, the embankment is to run straight across the bay into which the Bluther Burn falls, there being a sort of viaduct opposite the Bluther Burn to allow its water to get out to sea instead of being dammed up. After crossing the bay it hugs the shore, keeping within high-water mark till it reaches Low Valleyfield; then passing Culross it cuts in two the little town jetty that presently runs out to sea for boating purposes.[64]

. . . those using the road will have between them and the line a stretch of dead water at high tide, and at low water a miserably littered little bit of beach, never cleared by a strong rush in of water from the sea.[65]
. . . for about two miles the public will be shut out from the seashore and the open sea. I ask is this not grievous destruction of amenity and gross violation of public right?[66]

An alternative route to the north of the burgh, along higher ground and well away from the shore, was suggested by several correspondents: '. . . let the line be carried along the high ground to the back of Culross, and not along the shore, and the public will have no ground of complaint.'[67] The beauty of the northern shores of the Firth was surely just a tad exaggerated by one letter writer to *The Scotsman*, who declared:

> . . . the shore road before entering Culross is one of the most charming walks which can be found anywhere, and has frequently been held to be a miniature of that part of the Corniche Road leading between Nice and Mentone in the Western Riviera. Culross of late years has been much frequented by the artist and the antiquary, not to speak of the passing tourist and visitor; but should a hideous embankment (however low) be constructed along the shore, the burgh and district will simply be ruined as a summer resort.[68]

The writer concluded: 'One can only say it is a thousand pities if commercial prosperity can only be obtained by the sacrifice of amenity'.[69] Another correspondent was concerned about the safety of children at play:

> The inhabitants of Culross seem to forget that in permitting a railway to invade their delightful old town, they are killing the goose that lays the golden egg, and instead of a railway inducing summer visitors to come, they will, on the contrary, most assuredly select some more favoured spot, where they can enjoy a pleasanter view from their window than a railway embankment, and where their children can play upon the shore without danger from passing trains.[70]

Yet another loss to the railway was a village feature far from common in Scotland, as compared with its southern neighbour:

> To one who has resided all his life, as I have, in the neighbourhood, it is painful to contemplate the ruthless destruction of the most lovely part of the banks of the Forth and of the unique old burgh of Culross; as well as the cutting up of the village green, so uncommon in Scotland, the boast and the pride of Torryburn.[71]

In days long before political correctness, the prospect of dozens, more likely hundreds, of common, hard-drinking navigators descending upon this quiet nook of Fife was more than one correspondent could bear to contemplate: 'I trust this district . . . may long be spared from the inroad of the railway contractor and his "hordes of vandals" as represented by the navvies, with all their attendant operations so dear to the heart of the utilitarian.'[72]

It is interesting that some three years after the eventual opening of the line, it was reported, in an article entitled *The Town Council's Financial Dilemma*, that: 'the Council's claim against the North British Railway Company for the foreshore compulsorily acquired when the Dunfermline and Kincardine railway was constructed had not been settled yet. The Council claimed £1000, and the company offered £60.'[73]

The eventual outcome of this dispute, indicative of the shaft of broad daylight between the two parties, has been lost to the annals of history. Today, as a reminder of the events of over a century ago, some of the pools or lagoons that were created, trapped when the railway embankment was built across the beach, still remain, dammed on the landward side of the structure (Figure 1.6). These are now muddy hollows, partially drying out at low tide, with areas colonised by marsh plants. Yet again, a railway company – in this instance the North British – had despoiled an estuarine shoreline irrevocably.

In more highly populated coastal locations, anger at visual impediment, loss of access to the shore and loss of amenity became compounded by overpowering malodour and even the threat of disease.

Figure 1.6 Pool isolated by the railway embankment crossing the foreshore at Culross (photo: R. W. Duck).

Letters to the press were more voluminous and vociferous. Without adequate through drainage, railway embankments could act like dams impounding, in many cases fetid, polluted and therefore smelly waters. Ipswich lies at the head of the estuary of the River Orwell in Suffolk. At the tidal limit, at Stoke Bridge, a correspondent writing to *The Ipswich Journal* under the pen name of *Ambulator* noted with indignance:

> If there be one of your readers who has not experienced the annoyance, let him go there at low tide, and he will quicken his pace when he reaches the spot. There is another place in Stoke so bad that it is calculated to create a fever in the neighbourhood. I mean the ditch that leads from the Wherstead Road towards the Bathing place . . . that ditch used to empty into the Orwell, but the Railway people have been permitted to stop up the lower part of the ditch, and all the foul water from Station street and the houses for a long distance, runs into it, and remains in a stagnant state producing miasma, and endangering the lives of the inhabitants of all the cottages near the spot and it is sometimes in so fetid a state as to fill the rooms of cottages near with the stench, and renders the breathing difficult.[74]

The reference to 'Railway people' was a damming slight on those held responsible by *Ambulator* for this intolerable situation. Moreover, Ipswich was by no means an isolated example of the negative impact that railway embankments could have at the shore.

On the south coast of England at Southampton in Hampshire the docks were expanding rapidly during the second half of the nineteenth century, as were the railways. Today's traveller alighting at Southampton Central Railway Station is unlikely to realise that the street in which it is located – Blechynden Terrace – was once right at the waterfront, such is the extent of land claim, dock construction, road building and the subsequent infilling of disused docks that has taken place since the industrial revolution. Blechynden Station, an early predecessor to Southampton Central has long gone. From it, the London and South-Western Railway Company built a line to Dorchester which crossed the foreshore, known locally as 'mudlands', on a solid embankment. In 1848 the report of the Local Board of Health merely noted: 'The swamp produced by the embankment of the Dorchester Railway at Blechynden; this is the cause of a large amount of fever in the immediate neighbourhood.'[75]

The following year the correspondent *Olfactory* wrote to the *Hampshire Advertiser and Salisbury Guardian* alerting the reader to what many must have already been aware of, that:

In constructing the bank the shore has been improperly sunk in such a manner that the water is retained in very considerable pools – water of a most objectionable character, the water of a large drain which, flushed by the tide, deposits its contents with every ebb into these pools, polluting the air with its offensive effluvia.[76]

This situation continued until the autumn of 1895 when a decision was taken to fill in the pools with chalk rubble, which would also help to ease unemployment during the forthcoming winter months, in order to abate the intolerable nuisance, in particular the abominable stench that emanated from them:[77] 'Only the other day passengers in a train going to Bournemouth had to hold their handkerchiefs to their noses as they passed along.'[78]

Things were, however, a great deal worse at the opposite end of the country in the prosperous and rapidly expanding Scottish east coast port of Dundee.[79]

A coastal city transformed

Dundee was Scotland's original railway city and it was railways that were ultimately to change its waterfront character beyond all measure. The first line to make its mark upon the city was that from Newtyle to the north – the Dundee and Newtyle Railway was opened to traffic in 1831. This primary artery permitted the import of the vegetable produce of the fertile valley of Strathmore to feed the growing populace of Dundee. In turn, it was an export route for lime, coal, timber, grocery goods and, importantly, the 'essence of fertility', the 'produce' of the city's inhabitants, which would help to fuel agricultural output.[80] Although this line was famously tunnelled through the eastern flank of Dundee Law – the eroded remnant of a 400,000,000-year-old volcanic plug that dominates the city's skyline – and sliced through the streets of the western part of the city, it had no significant impact on the waterfront. Much was, however, to change as Dundee developed into a central railway hub with lines poised to impinge upon it from the remaining three quarters of the compass. The next railway line to be constructed was from Arbroath, opened in 1838 by the Dundee and Arbroath Railway Company. The line closely followed that of the Angus coast but its route into the eastern confines of the city was along an isolated embankment piled into the inter-tidal flats of the Tay, some 200–300 metres from the shoreline, reaching its eventual terminus near the city's Custom House.[81]

This mode of extension into the city might have been good for the railway company but it brought huge detriment to those who lived or worked in the area and those who had been accustomed to walking on the beaches or fishing or bathing – common recreational practices in this part of the city. The embankment effectively marooned the natural coastline along with the industries, jetties and piers that had developed along it during the late eighteenth and early nineteenth centuries. This was the whaling quarter of Dundee but now its piers and blubber boiling yards were cut off from the Tay. Some businesses relocated, others went into liquidation and, almost in an instant, this part of Dundee lost its ancient port character. The railway embankment was now delimiting and defining the waterfront. What had once been open water behind it was no longer of use to fishermen other than young boys trying their best to fish by hand line from the shore to landward of the railway-created offshore dam (Figure 1.7). Impounded in this way, the pool water was to become virtually stagnant and fetid, covered

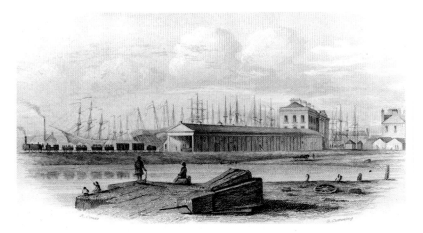

Figure 1.7 The embankment leading into the Dundee terminus of the Dundee and Arbroath Railway cut off the inter-tidal flats and beaches, in the foreground, from the Tay Estuary to create lagoons that collected the sewage and drainage from the east of the town (from *Forfarshire Illustrated*, 1843).

with green algal scum and in the summer months 'infested by a swarm of venomous insects.'[82] Proprietors and residents at the east end of Dundee petitioned the Town Council, the Harbour Trustees and the Railway Directors;[83] members of the public were appalled:

> But have the quick eye and the delicate taste of these philanthropists ever discovered the noxious effluvia arising from the stagnant mill ponds in some of our densely crowded localities? Instead of means having been taken to remedy these unhealthy places, more serious and alarming ones are daily arising up, apparently unheeded. I refer to what was formerly the sea beach between Trades' Lane and the Roodyards. This place is totally different now to what it once was, on account of its being wholly hemmed in by the railway embankments. It is a fact beyond dispute that this place is now the very receptacle of all the filth and contents of the common sewers of the eastern district of the town, and strange to say no adequate provision seems to be made for carrying it out into the river. The only place discoverable for this purpose is a small aperture across the railway, and to all appearance even this will soon be choked up, as the embanking of Dock Street is already within a few yards of it, and no appearances of preparations are making either for the outlet of stagnant water from this place, or inlet of fresh water from the river.[84]

It is thus hardly surprising that a noxious cocktail of raw sewage and, in the summer months, the refuse of whale blubber[85] that had previously discharged or soaked away directly into the Tay Estuary began to accumulate as there was little or no interchange with the main body of tidal water and any rudimentary culverts became choked. One of the pools of about 8 acres in surface area was observed to be connected to the Tay via only a single iron pipe.[86] Moreover, the pools also received the waste waters from various spinning mills.[87] Little wonder that the smell became unbearable and many Dundonians blamed the stagnant pools as a cause of disease as 'fever was raging'[88] in the town at that time:

> A few years ago, this place, with its salubrious air, beautiful banks, pebbled bottom and limpid waters, formed an agreeable and healthy retreat both to the philosopher and the humble artisan; but now it is converted into stagnant pools and deadly lakes. It must be apparent to everyone, that, from the vast quantities of putrid animal and vegetable matter daily deposited in this place, in a very short time its

pestilential miasmata will contaminate the atmosphere, and gener-
ate disease and death in this densely crowded neighbourhood in the
first instance, and will spread its ravages with the rapidity of lighten-
ing far and wide.[89]

Numerous people expressed their opinions regarding the loss of
amenity as a result of progress:

It has now become a matter of deep regret that those places along
our river side, which once afforded a quiet retreat to the contempla-
tive mind, as well as accommodation for sea bathing, are no more.
True, indeed, the geographical position of the places remains; but
the pebbled beaches, formerly washed by the 'rolling waves of
Neptune', are now transformed into unseemly nuisances covered
with all uncleanliness.[90]

A lecture at the city's Watt Institution drew attention to the high rate
of mortality arising from stagnant pools of water, 'an evil', as stated by
the speaker, a Mr Maxwell, 'very prevalent in Dundee.'[91] Furthermore
numerous calls were made for the infill of the pools. For instance:

These places ought immediately to be *filled up altogether*; or sufficient openings made for the free *ingress* and *egress* of fresh water,
so as to scour these pools properly at every tide from all dangerous
depositions which have been therein accumulating, whatever may be
the trouble and expense.[92]

A sub-committee of the Harbour Trustees was appointed to investigate
the matter and it recommended that the pool near to the terminus of
the Arbroath railway line should be filled up and 'the eye for letting in
fresh water to the second being put in better order'.[93] This eventually,
once it was agreed that this was to be at the expense of the Harbour
Trustees,[94] took place in stages resulting in the creation of more but
smaller pools, which progressively became concentrators of stagnant
effluent. Such land claim from the Tay Estuary was inevitable – these
hotspots of pollution and malodour simply had to be obliterated, with
proper drainage through the claimed land to permit sewage to exit into
the tidal waters of the Tay.

Dundee's third railway, which was to enter the city from Perth to the
west, opened in 1847. In the early 1840s there was no more fashionable

place to reside in the city but in a detached classical villa south-facing along the riverfront suburb of Magdalen Green. Between 1840 and 1841 the green had been levelled and drained to find work for unemployed artisans.[95] With unimpeded vistas of the Tay, wave-lapped beaches could be reached via steps from walled front gardens down gentle grassy slopes. The beaches were mainly pebbly, interspersed by low protruding headlands of basalt. The largest headland was originally known as Black Ness, the colour a reference to the hue of the basaltic lava that cropped out into the river. Hereabouts the foreshore was an area popular for walking, bathing, swimming, fishing, taking the air and for quiet contemplation. This was 'Arcadia on Tay',[96] as painted by George McGillivray in about 1840, showing schoolboys playing cricket on Magdalen Green. The glorious beaches of the Tay, from which other boys are fishing, are just a short boundary to seaward (Figures 1.8 and 1.9). Through time, Black Ness became known as Magdalen Point and the walk around the promontory along the 'Pleasure Walk Lately Made'[97] was the preferred form of Sunday afternoon recreation in the area. Paid for by public subscription, this was hailed as one of the most convenient and beautiful places of public recreation in Scotland.[98] The Dundee and Perth Railway Company, however, cared nothing about such prosperous rural idyll and proposed that their line should enter

Figure 1.8 *Magdalen Green before the Dundee to Perth Railway,* painted by George McGillivray (c. 1840). The view is eastwards along the Dundee waterfront on the northern shore of the Tay Estuary (image copyright Dundee City Council – Dundee's Art Gallery and Museums).

Figure 1.9 Approximately the same view depicted in Figure 1.8 as seen today (photo: R. W. Duck).

the city on an embankment to seaward of the magnificent beaches built through the inter-tidal flats – just as had happened less than 10 years previously in the east of the city. In this way three substantive pools would become isolated; a long, slender one beyond the western limit of the suburb, a shorter, but wider one immediately to the west of Magdalen Point and a third and very large pool beneath the city's Nethergate area. Such a mode of railway entry into Dundee from the west would maroon the beaches along with numerous waterfront jetties, disconnecting them from the Tay. Moreover, the view from Magdalen Green's salubrious villas would become impaired or even obscured from street level by an unsightly embankment. Members of the city's Guildry were seriously concerned about the proposal, noting that: 'One Railway had destroyed the privilege of bathing to the eastward of the town, and it would be a thousand pities that another should be allowed to do the same westward.'[99]

Not surprisingly, the local residents were incensed and called public meetings in massed opposition, mindful of the impact of the Arbroath to Dundee line.[100] The railway company, however, sought via the press to put the most positive of spins possible on the proposals:

Then it is said that the Magistrates of Dundee will not allow a railway to be carried through Magdalen Green, as it would destroy that beautiful park, which is the only place where the inhabitants can enjoy the privileges of recreation and sea-bathing. The promoters cannot believe that the Magistrates will oppose, as they (the promoters) intend to carry the Railway on the margin of the Tay for the entire length of the park – being about a mile – which it is obvious will give 'a finishing touch' to its beauty.[101]

A satirical virtue was even made of the potential loss of human life during the construction phase of the railway:

Another of the weak objections is, that carrying a Railway for a mile and a half through the streets of Dundee will kill a number of people. Of course it will; but instead of an objection, the promoters consider this one of the best features of the scheme, as the population of Dundee is increasing too fast.[102]

Perhaps mindful of the problems caused by inadequate drainage through the embankment of the Dundee and Arbroath Railway, the Directors of the Dundee and Perth Railway Company made it clear that they planned to secure the accommodation of bathers by means of arches through their proposed embankment and hinted that they might even establish a splendid esplanade and promenade by extending a structure westwards along the line of the railway.[103] The inhabitants noted, however, that they would be entirely excluded from the river to the west of Magdalen Green and that there was no proposal to construct arches in that quarter.[104] Despite the opposition, the influence of the railway company won the day. The only concession was that representations of the Town Council were successful in ensuring that ten culverts, 6 feet in height and 3 feet wide, would be constructed through the railway embankment for the purpose of carrying off the refuse of the western part of the town.[105] Thus, the Perth to Dundee railway embankment was constructed and the free draining, clean washed beaches of Magdalen Green were consigned to history.

By 1872, the esplanade to seaward of the railway embankment had been constructed with the western terminus of its masonry wall abutting the east-facing side of Magdalen Point, thereby claiming a further 44 acres of land from the Tay. Notwithstanding the construction of the esplanade and promenade, as in the east of the city, residents bemoaned the loss of their natural amenity:

I will never cease to regret that the abominable railway embankment was allowed to be made, and thus to deprive the inhabitants of such liberties, and the *ex adverso* proprietors from their valuable access to the river . . . What is the use of spending all the money in making those costly Esplanades and Pleasure Grounds, at present in course of extension west of the Magdalene [sic] Yard Point, on the river banks, if, in taking the air along them, one is met with a stench of sewage water at every turn! The air is completely impregnated with it.[106]

Because the beaches had been lost, Magdalen Point was almost the only spot remaining that afforded access to the Tay for swimmers and bathers in the area. Though it was popular, it gave access to dangerous currents and there were frequent occurrences of accidents and drowning.[107] Though the hinterland of Magdalen Green was less industrial and less densely populated than the eastern quarter of the city, there was a not surprising déjà vu, even with the more efficient drainage through the Perth to Dundee railway embankment:

I would . . . call attention to the state of the east of the Green, where the common sewer is still allowed to discharge its contents far inside the Railway embankment, instead of being carried beyond. This is a great nuisance to all who frequent the Green, and must be still more so to those residing in the neighbourhood.[108]

Now he would be a bold man, indeed, who would venture to disport his limbs in the muddy and filthy pools with which the railways have replaced the open beach.[109]

Dundee's fourth railway, borne by the ill-fated first Tay Railway Bridge about which so much has been written, the North British line from Burntisland, was to enter the city from the south. Its northern landfall was Magdalen Point, obscuring for ever what little then remained of the promontory beneath the tight eastward curve into the line's junction with the Perth to Dundee line. The last vestiges of the marooned beaches of Magdalen Green, like a giant bunker, are seen in photographs of the bridge's northern landfall (Figure 1.10), soon to be obscured for ever by infill and landscaping. The eventual fate of the remaining pools was inevitable and by 1878 they had been replaced largely by railway sidings and marshalling facilities. From east, west and south, in a period of only four decades, the ruthless endeavours of the railway companies had irrevocably transformed the waterfront of Dundee (Figure 1.11) – though

Figure 1.10 The first Tay Railway Bridge with a remnant of the Magdalen Green beach in the foreground cut off from the Tay Estuary (Image B14.056, courtesy of the Local History Centre, Dundee Central Library).

Figure 1.11 View westwards along Magdalen Green in Dundee; the Dundee to Perth railway line is to the south. High water mark was naturally at the break in slope of the green below the houses on the north side (photo: R. W. Duck).

it might today be perceived as 'ludicrously ideal',[110] all vestiges of its natural, rural character had been erased or obscured – it was now a city with a hard-engineered seaward edge.

Getting to the beach

All over Britain, wherever railways were built along the coastal edge, there were complaints that former rights of access either to the beach or along the coast had been taken away and this was an enormous source of frustration, even anger. The reaction to the coming of the railways to Dundee was typical – 'Why a few years ago . . . ', wrote 'An Elector in the Middle District' to the *Dundee Courier* in 1847:

> . . . Dundee possessed free and uninterrupted access to the river [the term often applied locally to the Tay Estuary], which was a never-failing source of great pleasure and of health to the inhabitants. This right has been allowed to be taken away by Act of Parliament with only a mere shadow of resistance, and without receiving a farthing of compensation.[111]

The situation was similar in the Edinburgh area. A bitter feud between members of the public and the North British Railway Company erupted in late 1860 when the latter completely closed off a much-used road to the beach. This coincided with the conversion of a horse-drawn branch line, originally opened in 1835, along the southern coast of the Firth of Forth between South Leith and Portobello[112] to steam locomotive power. What was then known as the Figgat (or Figgate) Whin Road, a right of way to the beach mid-way between the two settlements that had been in use for generations by pedestrians, horses, carts, gigs and even the cavalry, was completely blocked off at the seaward end by a wall, depriving the people of their long-standing rights of access.[113] Shortly afterwards, however, in 1863, legislation was introduced in Britain to ensure access to the shore under or across a railway. Specifically, Section 16 of *The Railway Clauses Act 1863* stated that:

> Where the railway cuts off access between the land and a tidal water or tidal lands, then and in every such case the company shall, during the construction of the railway, and from time to time thereafter, make, and shall permanently maintain, and allow to be used by

all persons, at all times, free of toll or other charge, all such footways and carriageways over, under, or across the railway, or on a level therewith, as the Board of Trade from time to time directs or approves: Provided always, as follows: –

(1) The company shall not be obliged to make a footway or carriageway over lands for the use of an owner or occupier who has agreed to receive and has been paid compensation for the severance thereof from the tidal water or tidal lands.
(2) The company shall not be obliged to make or to allow to be made a footway or carriageway in such manner as would interfere with the working or using of the railway.
(3) The expense of the making and maintenance of a footway or carriageway required to be made after the construction of the railway shall be defrayed by the persons or body interested in the tidal water or tidal lands for whose benefit or convenience the same is required.

Where the footway or carriageway is made across the railway on the level, then the manner of the making and watching of the level crossing shall be subject to the approval of the Board of Trade; and where the level crossing is made after the construction of the railway, then all expenses attending the watching thereof shall be defrayed by the persons or body interested in the tidal water or tidal lands for whose benefit or convenience the same is required.[114]

This legislation, however, appears to have had little or no impact on the numbers of complaints regarding the loss of access 'rights'. For instance, Broughty Ferry in Angus is situated some 6 kilometres to the east of Dundee on the northern shore of the outer Tay Estuary. Between here and neighbouring Monifieth the imposition of a railway embankment along the shore by the Dundee and Arbroath Railway Company in 1838 – part of what is today's East Coast Main Line – was still a matter of contention over three-and-a-half decades later. The correspondent 'Eastender' made a compelling case for securing access to the beach of the Tay, 'before it is finally closed from the public, as it has been done in our thriving town of Dundee.'[115] Indeed, the absence of appropriately sited under bridges, over bridges and level crossings was a continual source of irritation for local residents well into the late nineteenth century.

Nowhere was 'the law and custom of the neighbourhood' changed more than along the shore of the outer reaches of the Firth of Forth between Aberdour and Burntisland (Figures 1.12 and 1.13). A little over 3 kilometres apart, 'Fife's Railway Port',[116] a rapidly developing plethora of land claim, docks and sidings, was to be linked with its historic neighbour of Aberdour by railway along the approximately west-to-east trending rocky shore. This was part of the line to the Forth Railway Bridge via Inverkeithing, which opened in 1890. The route between Aberdour and Burntisland was, however, a popular footpath at the edge of craggy woodland known as The Heughs, a right of way beside spectacular outcrops of sedimentary and igneous rocks of Carboniferous age that had been trodden and enjoyed by local people and visitors alike since time immemorial. The tone of letters of complaint to the press was incandescent:

It seems that a railway is to take possession of the natural ledge on which the path runs through the woods; the rocks are to be blasted

Figure 1.12 The railway embankment provides Burntisland with an adjacent promenade but it severs the natural connection between the town's links and the beach (photo: R. W. Duck).

Figure 1.13 Pedestrian access to the beach at Burntisland is today focused through two underpasses, one of which is shown here. This exacerbates localised erosion, especially during the summer months, as trampling is concentrated into restricted areas (photo: R. W. Duck).

away, and the stones and rubbish to be shot onto the beach, and an embankment made all round the bay from the Silver Sands [of Aberdour] to Burntisland.[117]

The correspondent continued: 'My object in writing to you is to ask if the Right of Way Association cannot do something to stop this vandalism . . . '[118] This, along with other appeals[119] and deputations to the directors of the North British Railway,[120] fell, as was typical of the day, on deaf ears. The following year, the 'vandalism' began as blasting for the line commenced. It would appear that this was not a controlled process. On the evening of 5 September 1888, three men walking along Aberdour's High Street were witness to a very loud explosion followed by a shower of dislodged rocky debris over the village. The incident was so alarming that it aroused one of the three companions to write to *The Scotsman*:

I presume, from sheer instinct of preservation, we bolted, only to escape a shower of stones and tiles, &c; and on recovering somewhat from our fright, imagine our horror to see a man holding in his hands a piece of solid whinstone of fully twenty pounds weight, which must have dropped almost at our heels, a hole also being made in the roof of a house we were passing at the time, about two feet square.[121]

In fact this was apparently not an isolated incident and the local policeman was 'utterly helpless to prevent a repetition of these bombardments.'[122] The North British Railway Company had a lot to answer for: 'The destruction of the beauties of Aberdour through the formation of the railway is surely sufficient of itself without exactly trying to lay the houses in ruins.'[123] The woods and the rocky shore between Aberdour and Burntisland were now cut off from each other by the railway line along with a high masonry wall: 'With various cuttings and embankments, the line has run along on the site of the old favourite footway by the seaside, which, excepting at one or two points, it has practically obliterated.'[124] The prospect of a new footpath, to be constructed by the railway company on the seaward side of the railway, was a poor consolation. John Geddie, nearly four decades after the event, was somewhat philosophical about the coming of the railway to this part of Fife. He regretted the damage it had done to the landscape but realised that it was here to stay – in fact it is yet another section of today's East Coast Main Line – and that people must make the best of it:

Aberdour is still a 'sicht for sair een' [a sight for sore eyes]. Its great charm and virtue are in its situation. Its bay is a dimple of beauty in the stern lineaments of Fife; and lovely it remains, even in March weather, and after the railway has drawn an ugly scratch (Figure 1.14) across its wooded slopes.[125]

Today, the Fife Coastal Path has been squeezed in between the railway line and the shore for about half of the distance from Aberdour's Silversands Bay towards Burntisland; along the remaining section there was insufficient space for this to be contemplated.

Even as the twentieth century dawned there were still unresolved issues of access to the coast as a result of intervening railways. For instance on shore the Dee Estuary between Flint and Bagillt in Flintshire on the North Wales coast, a right of way and the only means of access to the Dee foreshore, was still being obstructed by the London and North-Western Railway Company's main line between Chester and Holyhead half a century after it had opened.[126] On the one hand railways had opened up the country and made previously undreamed of

Figure 1.14 The 'ugly scratch' of the Aberdour to Burntisland railway line today, as photographed in March weather looking across Silversands Bay. Further to the east the line completely obstructs access between the woods and the beach (photo: R. W. Duck).

opportunities for the traveller to visit coastal resorts and watering places. On the other hand, they had imposed barriers that were often resented by local people who no longer had free and unrestricted access to their beaches. Some of the complainers, however, were not simply concerned with recreation and quiet contemplation. They had ulterior motives for requiring access to the beach – in particular, access to the very materials *comprising* the beach – as will be explored further in Chapter 5.

And so the story goes on. More and more stretches of Britain's shores became changed for ever as the railways proliferated. Natural linkages between cliffs and beaches, between dunes and beaches were severed by great embankments and walls. Salt marshes were sliced through, damaged or destroyed. Natural processes of sediment transport were interfered with, impeded or stopped altogether and, as a result, coastal erosion often became more focused and forceful. Many of our coastlines were, however, not only to witness such changes, but also to experience tragedy, owing to the often precarious routes taken by the railways on the edge.

Notes

1. Matus, J. L. (2001), 'Trauma, memory, and railway disaster: The Dickensian connection', *Victorian Studies*, 43, 413–436.
   'Dreadful railway accident at Staplehurst', *The Times*, 10 June 1865.
2. Rich, F. H. (1865), South-Eastern Railway, *Report of The Secretary of the Railway Department, Board of Trade*, 41–44.
3. Letter from Charles Dickens to his doctor, Francis Charles Beard, Saturday 10 June 1865, available at http://www.railwaysarchive.co.uk/documents/Dickens_ Staple1865.pdf (last accessed 26 April 2014).
4. Matus, J. L (2001), 'Trauma, memory, and railway disaster'.
5. Flanders, J. (2012), *The Victorian City: Everyday Life in Dickens' London*, London: Atlantic Books.
6. Desecration of Edinburgh, *Glasgow Herald*, 2 June 1894.
7. Knowles, G. (1841), *Railroads. Observations on the Expediency of Making a line of Railroad from York to Scarborough*, Scarborough: C. R. Todd, 8 pp.
8. Knowles, G. (1841), *Railroads*.
9. Knowles, G. (1844), *Supplement*, Scarborough: C. R. Todd, 9 pp.
10. Corley, T. A. B. (2004), Rooke, John (1780–1856), *Oxford Dictionary of National Biography*, Oxford: Oxford University Press, available at http://www.oxforddnb. com/view/article/24061 (last accessed 26 April 2014).
11. Rooke, J. (1838), *Geology as a Science, Applied to the Reclamation of Land from the Sea, the Construction of Harbours, the Formation of Railroads, and the Discovery of Coal, with an Assumed Outline map of the Granite Formation of the Earth*, London: Ridgeway, 442 pp.
12. McGlashan, D. J., Duck, R. W. and Reid, C. T. (2009), 'Legal implications of mobile shorelines in Great Britain', *Area*, 41, 149–156.
13. McGlashan, D. J., Duck, R. W. and Reid, C. T. (2005), 'Defining the foreshore: coastal geomorphology and British laws', *Estuarine Coastal and Shelf Science*, 62, 183–192.
14. McGlashan, D. J., Duck, R. W. and Reid, C. T. (2005), 'Defining the foreshore: coastal geomorphology and British laws'.
15. Railway Clauses Consolidation Act 1845 c. 20 (Regnal. 8_and_9_Vict), available at http://www.legislation.gov.uk/ukpga/Vict/8-9/20/contents (last accessed 26 April 2014).
16. Railway Clauses Consolidation (Scotland) Act 1845 c. 33 (Regnal. 8_and_9_ Vict), available at http://www.legislation.gov.uk/ukpga/Vict/8-9/33/contents (last accessed 26 April 2014).
17. Railway Clauses Consolidation Act 1845.
18. Railway Clauses Consolidation (Scotland) Act 1845.
19. Railway Clauses Consolidation Act 1845.
    Railway Clauses Consolidation (Scotland) Act 1845.
20. 'Scottish Railway Notes: Scottish Landowners with Reference to Railways' (1848), *Herapath's Railway and Commercial Journal, Quarto Series*, Vol. X, No. 450, 22 January.
21. Anon. (1893), Obituary, Sir James Brunlees, *Minutes of the Proceedings of the Institution of Civil Engineers*, 111, 367–371.
22. Clarke, H. (1884), *A comparison of Morecambe Bay, Barrow-in-Furness, North*

Lancashire, West Cumberland &c., in 1836 and 1883, paper read before the Mechanical Section and the Section for Economic Science and Statistics of the British Association, September 1883, Southport, Barrow in Furness: Lord & Gill, 16 pp.

23. BBC News (24 March 2006), 'Man guilty of 21 cockling deaths', available at http://news.bbc.co.uk/1/hi/england/lancashire/4832454.stm (last accessed 26 April 2014).

24. Dixon-Gough, R. (2006), 'Changes in land use and their implications upon coastal regions: The case of Grange-over-Sands, Northwest England', in Dixon-Gough, R. W. and Bloch, P. C. (eds), *The Role of the State and Individual in Sustainable Land Management*, 14–31, Ashgate, Aldershot, Hampshire.
    Gilpin, L. R. (1997), *Grange-over-Sands: A Resort and its Railway*, Cumbrian Railways Association, Grange-over-Sands, Cumbria, 24 pp.
    Gilpin, L. R. (2008), *The Ulverstone and Lancaster Railway: The Challenge of Morecambe Bay*, Pinner, Middlesex: Cumbrian Railways Association, 95 pp.

25. 'Improvements to Grange', *The Lancaster Gazette and General Advertiser for Lancashire, Westmorland, Yorkshire, &c*, 22 August 1863.

26. 'Editorial, Grange-over-Sands', (1873), *The British Medical Journal*, 12 April, 412–413.

27. 'Editorial, Grange-over-Sands', (1873), *The British Medical Journal*, 12 April, 412–413.

28. 'Editorial, Grange-over-Sands', (1873), *The British Medical Journal*, 12 April, 412–413.

29. 'Editorial, Grange-over-Sands', (1873), *The British Medical Journal*, 12 April, 412–413.

30. 'Correspondence, Grange-over-Sands' (1873), *The British Medical Journal*, 26 April, 479.
    'Correspondence, Grange-over-Sands' (1873), *The British Medical Journal*, 3 May, 520.

31. 'Important Ratepayers Meeting at Grange', *The Lancaster Gazette*, 15 January 1876.

32. 'Important Ratepayers Meeting at Grange', *The Lancaster Gazette*, 15 January 1876.

33. '"Resorts" on Morecambe Bay – Grange-over-Sands', *The Lancaster Gazette*, 29 July 1876.

34. Gilpin, L. R. (1997), *Grange-over-Sands: A Resort and its Railway*.
    Gilpin, L. R. (2008), *The Ulverstone and Lancaster Railway*.

35. Clarke, H. (1884), *A comparison of Morecambe Bay, Barrow-in-Furness, North Lancashire, West Cumberland &c., in 1836 and 1883*.
    Adam, P. (1990), *Saltmarsh Ecology*, Cambridge: Cambridge University Press, 465 pp.

36. Dixon-Gough, R. (2006), 'Changes in land use and their implications upon coastal regions'.

37. 'The Opening of the Ulverstone & Lancaster Railway', *The Lancaster Gazette, and General Advertiser for Lancashire, Westmorland, Yorkshire, &c.*, 29 August 1857.

38. 'Railway landslip, Embankment washed away by the sea', *The Times*, 23 September 1918.

39. Ransom, P. J. G. (2001), *Snow, Flood and Tempest: Railways and Natural Disasters*, Hersham: Ian Allan Publishing, 176 pp.

40. Mason, D. C., Amin, M., Davenport, I. J., Flather, R. A., Robinson, G. J. and Smith, J. A. (1999), 'Measurement of recent intertidal sediment transport in Morecambe Bay using the Waterline Method', *Estuarine, Coastal and Shelf Science*, 49, 427–456.

41. Dixon-Gough, R. (2006), 'Changes in land use and their implications upon coastal regions'.
   Gray, A. J. (1974), 'The ecology of Morecambe Bay v. the salt marshes of Morecambe Bay', *Journal of Applied Ecology*, 9, 207–220.
   Gray, A. J. and Adam, P. (1974), 'The reclamation history of Morecambe Bay', *Nature in Lancashire*, 4, 13–20.
   'Blame the Railways for the Silting of the Bay', *The Westmorland Gazette*, 20 February 1998.
   Adam, P. (1990), *Saltmarsh Ecology*.

42. Friend, P. L., Velegrakis, A. F., Weatherston, P. D. and Collins, M. B. (2006), 'Sediment transport pathways in a dredged ria system, southwest England', *Estuarine, Coastal and Shelf Science*, 67, 491–502.
   Pirrie, D., Power, M. R., Wheeler, P. D., Cundy, A., Bridges, C. and Davey, G. (2002), 'Geochemical signature of historical mining: Fowey Estuary, Cornwall, UK', *Journal of Geochemical Exploration*, 76, 31–43.
   Lightfoot, P. (2011), *Exploring the Fowey Valley*, Penzance: Alison Hodge, 88 pp.

43. 'Cornish Railways', *The Royal Cornwall Gazette, Falmouth Packet, and General Advertiser*, 29 January 1864.

44. 'Cornish Railways', *The Royal Cornwall Gazette, Falmouth Packet, and General Advertiser*, 29 January 1864.

45. 'The Lostwithiel and Fowey Railway', *The Royal Cornwall Gazette, Falmouth Packet, and General Advertiser*, 22 October 1868.

46. 'New Railway between Fowey and Lostwithiel', *The Royal Cornwall Gazette, Falmouth Packet, Cornish Weekly News and General Advertiser*, 19 September 1895.

47. Fisher, S. (2012), *Rivers of Britain: Estuaries, Tideways, Havens, Lochs, Firths and Kyles*, London: Adlard Coles Nautical, 304 pp.

48. Teinbridge District Conservation Area Character Appraisals: Cockwood (undated), available at http://www.teignbridge.gov.uk/media/pdf/q/o/Cockwood_1.pdf (last accessed 26 April 2014).

49. Sekon, G. A. (1895), *A History of the Great Western Railway*, London: Digby, Long & Co., 373 pp.

50. Sekon, G. A. (1895), *A History of the Great Western Railway*.

51. 'Railways, South Devon', *The Morning Chronicle*, 21 October 1845.
   Sekon, G. A. (1895), *A History of the Great Western Railway*.
   Kay, P. (1990), *Rails along the Sea Wall*, Sheffield: Platform 5 Publishing Ltd, 60 pp.

52. 'The South Devon Railway', *The Royal Cornwall Gazette, Falmouth Packet and Plymouth Journal*, 4 December 1846.
   'The recent destructive gale, South Devon Railway', *The Morning Chronicle*, 28 October 1859.

53. Cadell, H. M. (1913), *The Story of the Forth*, Glasgow: James Maclehose and Sons, 298 pp.

Cadell, H. M. (1929), 'Land reclamation in the Forth Valley, I. Reclamation prior to 1840', *Scottish Geographical Magazine*, XLV, 7–22.

Cadell, H. M. (1929), 'Land reclamation in the Forth Valley, II. Later reclamation and the work of the Forth Conservancy Board', *Scottish Geographical Magazine*, XLV, 81–100.

54. Gostelow, T. P. and Browne, M. A. E. (1986), *Engineering Geology of the Upper Forth Estuary*, British Geological Survey Report, Vol. 16, No. 8, London: HMSO, 56 pp.

McLusky, D. S., Bryant, D. M. and Elliott, M. (1992), 'The impact of land-claim on macrobenthos, fish and shorebirds on the Forth Estuary, eastern Scotland', *Aquatic Conservation: Marine and Freshwater Ecosystems*, 2, 211–222.

55. Undiscovered Scotland: Culross, available at http://www.undiscoveredscotland. co.uk/culross/culross/index.html (last accessed 26 April 2014).

56. Whatley, C. A. (1987), *The Scottish Salt Industry 1570–1850 – An Economic and Social History*, Aberdeen: Aberdeen University Press, 169 pp.

57. Geddie, J. (1894), *The Fringes of Fife*, Edinburgh: W. & R. Chambers Ltd.

58. Geddie, J. (1894), *The Fringes of Fife*.

59. Geddie, J. (1927), *The Fringes of Fife: New and Enlarged Edition,* Edinburgh: W. & R. Chambers Ltd.

60. Adamson, D. (2008), 'A coal mine in the sea: Culross and the Moat Pit', *Scottish Archaeological Journal*, 30, 161–199.

Graham, A. (1969), 'Archaeological notes on some harbours in eastern Scotland', *Proceedings of the Society of Antiquaries of Scotland*, 101, 200–285.

61. Geddie, J. (1927), *The Fringes of Fife: New and Enlarged Edition*.

62. Letter, *The Scotsman*, 7 February 1898.

63. 'Dunfermline and Kincardine Railway Bill', *The Scotsman*, 4 February 1898.

64. Letter, *The Scotsman*, 7 February 1898.

65. Letter, *The Scotsman*, 7 February 1898.

66. Letter, *The Scotsman*, 7 February 1898.

67. Letter, *The Scotsman*, 7 February 1898.

68. Letter, *The Scotsman*, 8 February 1898.

69. Letter, *The Scotsman*, 8 February 1898.

70. Letter, *The Scotsman*, 8 February 1898.

71. Letter, *The Scotsman*, 8 February 1898.

72. Letter, *The Scotsman*, 10 February 1898.

73. 'Lord Bruce and Culross foreshore: The Town Council's financial dilemma', *The Scotsman*, 22 March 1909.

74. 'The Public Health', *The Ipswich Journal*, 12 September 1857.

75. 'Report of the Local Board of Health', *Hampshire Advertiser and Salisbury Guardian*, 11 November 1848.

76. 'The Blechynden Nuisance', *Hampshire Advertiser and Salisbury Guardian*, 11 August 1849.

77. 'Southampton County Council, The Western Shore', *Hampshire Advertiser*, 12 October 1895.

78. 'Southampton County Council, The Western Shore – The Nuisance to be Abated', *Hampshire Advertiser*, 21 September 1895.

79. Duck, R. W. and McKean, C. (2011), 'Docks, railways or institutions: Competing images for mid-nineteenth century Dundee', in Whatley, C. A., Harris, B. and

Miskell, L. (eds), *Victorian Dundee: Image and Realities*, 151–172, Dundee: Dundee University Press.

80. Anon. (1833), 'Account of the Dundee and Newtyle Railway drawn up from official documents', *The Quarterly Journal of Agriculture*, 4, 1–29.

81. Cumming, G. (1843), *Forfarshire Illustrated: Being Views of Gentlemen's Seats, Antiquities, and Scenery in Forfarshire with Descriptive and Historical Notices*, Dundee: Gershom Cumming, 140 pp.
    Thompson, J. H. and Ritchie, G. G. (1930), *Dundee Harbour Trust Centenary 1830–1930: History and Development of the Harbour of Dundee*, Dundee: Dundee Harbour Trust.

82. 'The monster nuisance', *Dundee Courier*, 3 July 1850.

83. 'Town Council Proceedings', *Dundee Courier*, 18 February 1845.

84. Letter, *Dundee Courier*, 21 January 1845.

85. Letter, *Dundee Courier*, 26 August 1845.

86. Letter, *Dundee Courier*, 26 August 1845.

87. Letter, *Dundee Courier*, 26 August 1845.

88. Letter, *Dundee Courier*, 26 August 1845.
    'Town Council Proceedings', *Dundee Courier*, 10 March 1846.

89. Letter, *Dundee Courier*, 21 January 1845.

90. 'Dock Street nuisance', *Dundee Courier*, 10 August 1847.

91. 'Sanitary state of towns', *Dundee Courier*, 27 January 1846.

92. Letter, *Dundee Courier*, 26 August 1845.

93. 'Harbour Trustees', *Dundee Courier*, 19 August 1845.

94. 'Harbour Trustees', *Dundee Courier*, 30 September 1845.
    'Harbour Trustees', *Dundee Courier*, 2 December 1845.

95. Barrie, D. (1890), *The City of Dundee Illustrated: Reminiscences and Remarks, Critical and Otherwise, relating to Dundee and Neighbourhood, and to Certain Events Therein and Elsewhere in Scotland during the last Sixty Years, and relating to Local Government in Scotland*, Dundee: Winter, Duncan & Co.

96. McKean, C and Whatley, P. with Baxter, K. (2008), *Lost Dundee: Dundee's Lost Architectural Heritage*, Edinburgh: Birlinn.

97. Map of Magdalene Yard and Lands Adjoining, by David Neave, Architect, Dundee, 29 June 1813.

98. Harris, B. (2009), 'Merchants, the middling sort, and cultural life in Georgian Dundee', in McKean, C., Harris, B. and Whatley, C. A. (eds), *Dundee: Renaissance to Enlightenment*, 243–267, Dundee: Dundee University Press.

99. 'Guildry Incorporation: Dundee and Perth Railway', *Dundee Courier*, 14 January 1845.

100. 'Public meeting of the inhabitants: Dundee and Perth Railway', *Dundee Courier*, 21 January 1845; 18 February 1845.

101. 'Dundee and Perth Railway Second Supplementary Prospectus', *Dundee Courier*, 3 December 1844.

102. 'Dundee and Perth Railway Second Supplementary Prospectus', *Dundee Courier*, 3 December 1844.

103. 'Guildry Incorporation: Dundee and Perth Railway', *Dundee Courier*, 14 January 1845.

104. 'Public Meeting of the inhabitants: Dundee and Perth Railway', *Dundee Courier*, 21 January 1845.

105. 'Town Council Proceedings, Dundee and Perth Railway Bill', *Dundee Courier*, 3 June 1845.

106. Barrie, D. (1890), *The City of Dundee Illustrated*.

107. 'A swimming bath for the million', *Dundee Courier*, 17 May 1861.
'Sad case of drowning at Magdalen Point', *Dundee Courier*, 27 July 1868.
'Dundee Police Commission', *Dundee Courier*, 9 October 1868.
'Exciting scene at the Magdalen Point', *Dundee Courier*, 29 July 1872.
'Narrow escape of two boys from drowning', *Dundee Courier*, 30 July 1872.
'Sad case of drowning at the Magdalen Green', *Dundee Courier*, 22 July 1876.

108. 'The Magdalen Green', *Dundee Courier*, 28 January 1857.

109. 'A swimming bath for the million', *Dundee Courier*, 4 May 1861.

110. Quotation by broadcaster, actor and former Rector of the University of Dundee, Stephen Fry, in Dundee Waterfront, available at http://www.dundeewaterfront. com (last accessed 26 April 2014).

111. Letter, *Dundee Courier*, 27 October 1847.

112. 'Leith', *The Scotsman*, 21 September 1833.
'Leith Branch Railway', *Caledonian Mercury*, 24 November 1834.
'Opening of the Leith Branch Railway', *The Scotsman*, 11 March 1835.
'Leith', *The Scotsman*, 18 March 1835.

113. 'The Figgat Whin Road – Right of way meeting', *The Scotsman*, 12 November 1860.
'The Figgat Whin Road', *The Caledonian Mercury*, 5 January 1861.
'Right of way meeting: Shutting up of the Figgat Whin Road', *The Scotsman*, 5 January 1861.
'The Craigentinny right of way case', *The Scotsman*, 31 July 1861.

114. Railway Clauses Act 1863 c. 92 (Regnal. 26_ and_27 Vict), available at http:// www.legislation.gov.uk/ukpga/Vict/26-27/92/contents (last accessed 26 April 2014).

115. 'Railway crossings – Broughty Ferry', *The Dundee Courier and Advertiser*, 23 January 1874.

116. Marshall, P. (2001), *Burntisland: Fife's Railway Port*, Usk: The Oakwood Press, 192, pp.

117. 'Aberdour path – the Railway Company and right of way', *The Scotsman*, 28 April 1887.

118. 'Aberdour path – the Railway Company and right of way', *The Scotsman*, 28 April 1887.

119. 'Rights-of-way', *The Scotsman*, 16 May 1890.

120. 'Aberdour', *The Scotsman*, 27 May 1887.

121. 'Blasting at Aberdour railway', *The Scotsman*, 8 September 1888.

122. 'Blasting at Aberdour railway', *The Scotsman*, 8 September 1888.

123. 'Blasting at Aberdour railway', *The Scotsman*, 8 September 1888.

124. 'Railway extension in Fife', *The Scotsman*, 31 August 1888.

125. Geddie, J. (1927), *The Fringes of Fife: New and Enlarged Edition*.

126. 'Access to the Dee foreshore', *Cheshire Observer*, 19 May 1900.

2

Over the Edge

Railway accidents and grouse come into season about the same time.
The Saturday Review, 8 August 1863

Off the rails

At around 10.56 a.m. on Monday, 22 October 1979, the 9.35 a.m. Glasgow to Aberdeen express passenger train crashed at a speed estimated at about 60 miles per hour into the rear end of the disabled 8.44 a.m. Glasgow to Dundee local service. Five people died and fifty-one were injured. The accident took place alongside the edge of Invergowrie Bay, right at the very spot where the Perth to Dundee railway line hugs the northern shore of the Tay Estuary most closely. The force of the collision threw the rearmost four coaches of the five-coach front train over the low, inclined, masonry sea wall and into the Tay below. Fortuitously it was low tide at the time.[1] Accidents such as this were, mercifully, rare by the closing decades of the twentieth century.

By contrast, nineteenth-century railway travel in Britain was often a very dangerous affair. There are numerous accounts of often serious accidents due to derailments and collisions. Annual accident numbers reached a peak in the 1870s[2] as winning the race to build and operate lines was often seen as more important than standards of construction and safety of operation. Indeed so frequent were railway accidents that, during the early 1850s, the London Sunday newspaper, *The Examiner*, carried a column headlined *The Railway Accidents of the Week*. In other newspapers, headlines were often blasé: *Another Railway Accident* was used dismissively on quite literally hundreds of occasions,[3] with the addition of adjectives such as *Fearful, Frightful, Terrible, Dreadful* or *Appalling*, as appropriate. As standards of construction, maintenance of track and rolling stock, signalling and general safety improved, so the frequency of accidents became reduced.

However, even in the 1890s, well after the peak of railway mania, there were particularly black weeks, such as the last week of August 1891, perhaps lending some credence to the glib concept of 'The Railway Accident Season'.[4] Reporting by *The Aberdeen Weekly Journal* was typical of the period; its edition of 2 September 1891 containing no less than five headlines that collectively paint the grim story of a truly calamitous week across the network: *Disastrous Railway Accidents in England* (a reference to two separate accidents on the Lancashire and Yorkshire Railway); *Alarming Railway Collision on the Leicester Railway*; *Serious Railway Accident at Ramsgate*; *Three Trains Wrecked* (a collision involving no less than three goods trains at Retford in Nottinghamshire) and finally, *Peculiar Railway Accident near Leeds*.[5] Of these incidents, however, only the one at Ramsgate, Kent, had a coastal setting. Railway embankments have always been potentially hazardous, leading to accidents involving structural collapse, subsidence or locomotives becoming derailed, plunging from the line and dragging their load with them. At the coast, however, there is an added dimension, an extra element of hazard, and railway lines perched right at the edge of our shores have, perhaps not surprisingly, been sites of many remarkable disasters over the decades.

Immediately to the south east of the long-closed Mound station, the single track of the Highland Railway between Rogart and Golspie in Sutherland skirts the inner shores of Loch Fleet, an arm of the mighty Dornoch Firth. An alarming accident took place, close to this once rural halt, on 26 November 1885. An obsolete sight today, a mixed train of eighteen vehicles – passenger coaches and goods wagons combined – had left Golspie at half-past four en route for Inverness to the south. As it approached Mound station, at the point where the railway formation is closest to the water's edge, the leading wagon next to the engine suddenly left the rails and toppled into the sea, taking with it fourteen carriages with passengers and the other wagons. All fell over the embankment and into Loch Fleet. Fortunately, the driver had been slowing down the engine at the time of the accident and the passengers, although very much shaken and bruised, all luckily escaped with their lives.[6] It was said that the passengers 'were all stunned by the concussion, and the water at this point being between 2 and 3 feet deep, all of them suffered from immersion.'[7] Since it was a cold November day, it was speculated that the immersion might have serious after effects. In total, however, six persons were injured and the cause of the accident was believed to have been the snapping of a wagon axle.[8] Most likely because there were no fatalities on this occasion and the remote location is far from any metropolis, this remarkable accident received very little coverage in the British press. The same can also be said for a similar accident in South Wales two decades earlier.

Prior to 1964 – a victim of Dr Beeching's cuts – a short branch of the Great Western Railway from Johnston to Neyland in Pembrokeshire hugged the eastern shore of the tidal creek of Westfield Pill, an arm of Milford Haven. Brunel had engineered this line, which opened in 1856, as a gateway to Ireland via the deep water port of New Milford,[9] variously and latterly known at Neyland. Today the creek is a nature reserve and the old railway formation carries the 'Brunel Cycle Track'. Ten years after the line opened it was the site of a 'frightful railway accident'.[10] A train had left Johnstone at about 7.00 p.m. on 18 August 1866. It was running about 22 minutes late for the final leg of its 7-kilometre journey, a gentle decline into New Milford. Less than a kilometre north of the terminus, a third class carriage coupled to the tender left the rails whilst traversing a curve. One of the passengers in this carriage attracted the driver's attention by calling out. The driver whistled for the brakes and reversed the engine. However, this action broke the coupling between the carriage and the tender.[11] The carriage then left the rails, swerved across the line, dragging with it a second class carriage and a guard's van. The three vehicles became separated from each other, cascaded down the 3-metre-high bank and were strewn into the creek, where they came to rest embedded in the mud. Of the four other carriages comprising the train, all remained upright but two were derailed. Fortunately, and not for the first time in such an accident, it was low tide at the time. Here the maximum tidal range between low and high water is large; over 7 metres (24 feet). Had the tide been in at the time, the occurrence could have been tragic. As it was, some eight passengers were reported to have been slightly injured. Excessive speed, owing to the train running late, and a defective coupling were determined to have been the cause of the accident.[12]

A week after the derailment, a gallant passenger on the train described his experience in a newspaper article entitled 'A Tour to Tenby'. His wonderfully florid prose brings a sense of romance to the scene that epitomises narratives of that day. No doubt, he was able to dine out on his story on many occasions:

> It was between six and seven o'clock when the train left Haverfordwest for Neyland Ferry. Shortly after passing Rosemarket

Station, a tributary tidal stream, which flows into Milford Haven, was seen on our left hand. The evening light seemed gradually thickening. I looked out of the carriage windows. On our right, the precipitous face of the rock, with its angular projecting masses, was richly festooned with tresses of luxuriant vegetation – weeds and foliage forming the tapestry which Nature was silently weaving for the partially uncovered wall.

On the left hand, the river was gradually rolling along its bed for the night. All seemed secure and tranquil – the train like the stream seemed fulfilling its part of steady progress; but suddenly the pace of the former fearfully increased – a noise as of the wheels grinding something beneath the carriage was heard, and the carriage swayed unsteadily from side to side. Suddenly we came to a dead stop. Four of us were in the compartment of a first-class carriage. I was sitting with my face looking in the direction of our journey, on the side near the rocky boundary of the rail. At the other end of the carriage, on the same side, a lady occupied the corner-seat. Suddenly she rose with an exclamation of terror from her place, and rushed to the door next to me, motioning to be let out; for she was speechless with fright. *What is the matter?* I said, *Let us remain in the carriage*; though I expected every moment to find it hurled over, or its sides burst in by some force outside. The lady pointed to the window, and then I saw a carriage lying head foremost at the bottom of the steep bank, half in the river; a second carriage turned on one side parallel with the river, its wheels lifted up in the air; and a luggage van lying down the bank, with its front part close to the water. But I looked and listened in vain for the engine and tender – they were gone a-head [sic] somewhere.

The lady saw the carriages run off the line and down the bank, and it was this spectacle which shocked her, and drove her for the moment distracted. Waiting for a brief space to see the carriage steady, we got out; and then such a confused uproar of agonised shrieks and moanings came from the overturned carriages, as both alarmed and distressed those who were looking on unhurt and unshaken. The sloping bank was composed of chippings from the rock, and larger stones, and afforded only an unsafe footing. At first everybody was paralyzed, and the guard of the train was pallid with emotion. All dreaded to approach the carriages, expecting to see dead and mangled bodies requiring to be lifted up to the line. A few of the male passengers stepped down the bank, and the doors of the upset carriages were opened; and one by one the unfortunate occupants were lifted out. A mother with an infant in her arms expecting instant death, cried out, *Save my child*, and held it out to her rescuer, regardless of her own position. An aged woman, trembling all over with fright was lifted up to the bank, and so, one by one, the passengers were landed, and it was found that not one life was lost, and not a bone was broken! So critical was the position of the people in the carriage first hurled down the bank, that had the accident happened half-an-hour later they would have been all engulphed [sic] in the tide – for before we moved from the spot the first compartment was completely under water.[13]

In drawing his piece to a conclusion, the fortunate and relieved passenger brought a sense of Victorian perspective to that August evening's events on the edge of Pembrokeshire:

After the escape from such an accident, I need not detail the subsequent *désagrémens* of the journey. We were due in Tenby at 8.11 – we were not there till 11.0. Our lodgings were closed for the night – the hotel we were referred to did not contain a spare bed – I expected to walk the streets; but, fortunately, *the Lion* found us shelter, and there, wearied, worn-out, and with visions of the carriages hurled off the line and smashed, I sank into troubled slumbers.[14]

Here, as at Mound, the sea had played no direct role in causing the accident but a railway embankment right at the water's edge was clearly a potentially dangerous spot in nineteenth-century Britain. A few years earlier, on the peaceful, summer Sunday afternoon of 8 July 1860, a terrible and rather similar accident had taken place at Granton on the shores of the Firth of Forth near Edinburgh. A train left the tracks whilst passing along the massive embankment that had been built along the shore, a structure referred to in the previous chapter. At first it was rumoured that the train in question was a passenger train of the Edinburgh, Perth and Dundee Railway that had left Edinburgh at 4.30 p.m. However, the accident had, in fact, occurred to the engine and tender while they were returning to the locomotive depot at Scotland Street, after having safely conveyed the passenger train to Granton. The engine had just left Granton, at about five o'clock, and was proceeding rapidly along the high embankment, when:

... from some unknown cause the engine and tender went off the rails at a point opposite Wardie Cottages, a little to the westward of

Wardie Hotel, and dashed over the embankment into the sea, a distance of about thirty feet, carrying with them the railing of a bridge
which at that point crosses the road and a portion of the low parapet
wall skirting the line.[15]

Six persons were on the footplate at the time. The engine driver, his
son, his brother-in-law and a pointsman were killed on the spot and
their bodies were heavily mutilated. The stoker made a miraculous
escape, with only a dislocated shoulder, by jumping off the engine
before it went over the embankment. Even more incredible was the
survival of the porter. He was propelled over the embankment and fell
down the masonry bulwark close to the sea; it was almost high water at
the time. He was severely bruised by the fall, suffered cuts to the head,
was scalded about the face by the steam from the engine but was able to
walk to his home at Wardie Square, opposite the scene of the accident.
A family, who had been relaxing in the late afternoon sunshine, were
also injured. Sitting reading on the stones at the foot of the embankment, husband, wife and two young children, one an infant, were all
severely scalded by the steam from the engine as it toppled. The injured
porter's child, who was also playing near the water's edge close to the
family foursome, fell into the water, supposedly from fright, but was
rescued and taken home unhurt.[16] As the tide receded, the engine
could be seen lying bottom upwards, while the tender lay close behind
it on its side. The tragedy, in contrast with events at Mound and New
Milford – since fatalities were involved and it occurred so close to
Scotland's capital – received widespread newspaper coverage and even
reached the pages of the *London Illustrated News*,[17] which published an
engraving of the aftermath (Figure 2.1).

Serious though this accident was, it would have been an even
greater tragedy had the earlier passenger service been the one to fall.
Today this railway has gone and the uppermost part of the high and
formerly intrusive embankment from which the engine fell has been
demolished. It is noteworthy that, yet again the sea played no direct
role in causing this tragedy but in this case it certainly contributed to
it. The Mound, New Milford and Granton 'incidents', however, were
no mere 'one-off' occurrences – railway lines built along the edge of
our coasts have witnessed a curious variety of accidents over the years
whereby engines, tenders and coaches have either toppled into the sea
or teetered on the brink of collapse. Moreover, the sea has not always
played such a passive role.

Figure 2.1 Engraving of the aftermath of the Granton railway accident (from *The Illustrated London News*, 21 July 1860).

Calamity at the Kent coast

An innocent bystander, a hawker named Jabez Grainger, was the unfortunate sole fatality at Ramsgate, the harbour terminus of the London, Chatham and Dover Railway. He was standing at his barrow serving customers on the dark, very wet and thoroughly miserable evening of the last day of August 1891. Without warning a locomotive plunged through a very thick retaining wall at the end of Ramsgate station and onto the roadway below.[18] Four other people were seriously injured, being buried in the debris that resulted from the impact. The engine, running tender first and pulling an empty excursion train, came into Ramsgate from Margate for the purpose of taking on passengers.[19] Descending an incline to the station, the train failed to stop at the platform end, knocked down the buffers, ran across the public road, killing Grainger, and propelled itself onto the sandy beach. Of the thirteen carriages, one was completely smashed on the roadway; three were overturned and seriously damaged, whilst the rest remained on the rails. The jury at the subsequent inquiry returned a verdict of accidental death, noting that, as a similar accident had occurred *twice* previously at the station, there was a need to make alterations to the approach to the terminus.[20] It is some small comfort that the accident took place on a night of thoroughly inclement weather when most people were indoors. Had the evening been warm and dry, the roadway and adjacent beach would likely have been crowded.

Forty years earlier, an incline had contributed to another serious accident along the coast at Folkestone. On this occasion, however, no lives were lost. When the South-Eastern Railway, rival to the London, Chatham and Dover, arrived close to the port in 1843, a reporter from *The Times* rejoiced that:

> If any one had told you a few months or even a few weeks ago, that in this present month of June you should rise from your bed in London, breakfast on the sea-shore in England, lunch in Boulogne, spend a few hours there and have a good laugh at the Napoleon monument, and yet that you should be back in London again at such an hour, that (if you were a man of fashion and had a patient stomach) your dinner would just be ready for you – if any man had told you this, probably you would have set him down as a Bedlamite, or the projector of some Joint-Stock Aerial Ship Company.[21]

The intrepid traveller could undertake this mission in about 16 hours, allowing a little over 2 hours in Boulogne. However, an hour-and-a-quarter were lost transferring to and from the railway station – located on the cliff top above the town – and the harbour.[22] The connection time was, however, to become reduced greatly just a few years later when a short branch was built linking the harbour with the main line. This descended the town's East Cliff; in elevation by 111 feet over a length of just 1,328 yards.[23] The branch was constructed right across the harbour, over a series of brick arches, dividing it into inner and outer basins that were eventually inter-connected via a swing bridge (Figure 2.2). The harbour had been designed by the consulting engineer Thomas Telford, its construction at the site chosen being aided by the presence of a spit of gravel, projecting to the north east, that helped to create the desired shape.[24] The harbour station was originally located on a piece of reclaimed coastal shingle and the end of what was then known as the West Pier, which had been built in 1810 and acquired by the South-Eastern Railway Company for £18,000 after the Folkestone

Figure 2.2 The branch railway to Folkestone Harbour Station crossing the harbour over a series of brick arches (photo: R. W. Duck).

Harbour Company went bankrupt in 1842.[25] The reason for the latter's demise was due simply to the interruption of natural coastal processes; the pier was acting like a large groyne. As such, it was trapping the sediment being transported along the Channel coast by the prevailing south westerly winds, which drive longshore drift from west to east.[26] Thus a bank of gravel had built up on the western side of the pier, preventing the access of vessels and ultimately causing the company to crash. The railway company, however, in order to operate its cross-Channel steamer route to and from Boulogne, carried out successive extensions to the pier in order to maintain deep water berths as more and more sediment accumulated. Thus, it became known as the Promenade Pier in 1863, the New Pier in 1883 and had its final extension in 1905.[27] No longer 'new', today it is simply referred to as Folkestone Harbour Pier (Figure 2.3).

Today, Folkestone Harbour Station is a sad and forlorn place. Derelict, hollow, weed-infested and decaying, its rusting and partially uplifted rails have not seen a train since 2009, though an association has been formed with the aim of preserving it as a heritage route known

Figure 2.3 Aerial image of Folkestone, Kent. Note Folkestone Harbour Pier and the railway branching towards it from the main line dividing the harbour into inner and outer basins. The accumulation of sediment, much of it now built over, on the south-western side of the pier is striking (©2014 Google – Cnes/Spot Image, DigitalGlobe, Getmapping plc, Infoterra Ltd & Bluesky, Landsat – Imagery July 2013).

as 'The Remembrance Line'.[28] An eerie sight between the platforms is a five-headed statue, a group huddled together on a carpet that drapes across the rusting rails (Figure 2.4). *Rug People*, the creation of Paloma Varga Weisz,[29] was inspired by the illustrious history of this terminus; a railway that brought First World War soldiers to the harbour to embark for France, as well as in its final years being used by occasional Venice Simplon Orient Express services. The desolate, curving western platform is the only means of pedestrian access to the equally forlorn 'New Pier', the once bustling terminus of the Boulogne ferry service, prior to its Channel Tunnel-induced demise.

On Tuesday, 12 August 1851 a special train – the *Glorious Twelfth* – started from Paris, filled with 286 passengers and their luggage, bound for The Great Exhibition at Crystal Palace in London. On arriving at Boulogne the train was running half an hour behind schedule. Although the steamer – the *Lord Warden* – made a good passage across the English Channel, great delay occurred on arrival at Folkestone

Figure 2.4 'Rug People' between the platforms of the decaying Folkestone Harbour Railway Station on the Harbour Pier (photo: R. W. Duck).

Harbour Station.[30] The luggage was offloaded onto the platform for inspection by the customs officials. However, the quantities of bags and cases deposited upon the platform far exceeded the expectations of the railway attendants. Checking the luggage and loading the various items into the appropriate carriages led to further delays so that the train, postponed until 3.00 p.m., did not depart for London until 45 minutes later. On its ascent, up the steep and curving branch line, around half-way up the cliff, the coupling between the locomotive and the leading carriage gave way and the whole of the train was precipitated back down the steep incline towards the harbour; 'the velocity increasing alarmingly, as the breaks [sic] were utterly inadequate in such an emergency.'[31] Onlookers thought that the whole train, comprising nine carriages, would plunge into the sea but workmen and attendants quickly threw planks of wood and beams that were to hand onto the line in front of the descending carriages. These served to reduce the momentum of the train, certainly lessened the aftermath and 'substituted a moderate concussion for a violent one.'[32] On reaching what was then the end of the line, the rear carriage smashed through the fixed buffers with such force that its upper body became shorn clean off from the undercarriage and was projected through the wooden end of the station and onto the pebbly shore to the west of the pier some 6–7 feet below, leaving the damaged wheels of the van on the rails.[33] Luckily, this contained items of luggage and no passengers; moreover, yet again, the tide was out at the time. None of the other carriages of the train appeared to have suffered any injury. Several passengers sustained cuts and bruises but remarkably there were no serious injuries. It is also remarkable that – despite all the chaos and confusion – the local workforce re-formed the somewhat shortened train and it was able to re-commence its journey to London, arriving just 3.5 hours behind schedule.[34] Whilst this was taking place, any wreckage cleared and the luggage recovered, the passengers had to wait on the platform where 'a very tumultuous scene presented itself.'[35] Newspapers reported their relief in relation to the racial stereotypes of those on board, all of whom must have realised that this could so easily have been a major tragedy but mercifully no lives had been lost:

The Frenchmen who had not fainted or sat down crying, congregated, exciting each other, and suggesting all kinds of impracticable or idle expedients to prevent a recurrence of the accident. It was their general impression that English railways are so organised as to create accidents and that this was got up to receive them in England. The Spaniards and Italians remained firm, and some of the English gentlemen, seeing that there was no want of their assistance on the train, used their best exertions in reassuring their French fellow-passengers and attending to the ladies.[36]

This was not, however, the last time that the South-Eastern Railway Company's progressively lengthening pier was to be implicated with an accident at the coastal edge, but of a very different nature to the one described above.

A bastard shaly rock

Many coastal towns aspired to become watering places in the nineteenth century and Pwllheli on the Llŷn Peninsula of north-west Wales was no exception. To that end, a correspondent in the *North Wales Chronicle* some seven years before its eventual arrival noted that: 'Pwllheli has hitherto been comparatively neglected by tourists and pleasure seekers' and '. . . a railway is wanting.'[37] 'The railway, however, will produce a wonderful revolution in this particular [town], and to no town that we are aware of are its beneficial effects more likely to be felt than in Pwllheli.'[38] And in due course, in 1867, the railway did arrive in this 'salt water pit': 'The man who christened this charming and snug little Welsh town "Pwllheli", was a stupid old dotard, with no aptness for names, nor any appreciation of the cosily picturesque . . . '[39] With Pwllheli as its northern terminus, the railway along the coast from Machynlleth did indeed transform the market town into a fashionable watering place. In 1894 the great Victorian entrepreneur and developer Solomon Andrews began to build a horse-drawn tramway, primarily with the aim of transporting stone.[40] Andrews had opened up a quarry some 4 kilometres south westwards of Pwllheli, along the coast of the Llŷn Peninsula. Stone was integral to his plans to build a promenade, hotel and fine houses in the watering place. The tramway, however, soon began to develop into a tourist attraction in its own right, so a few months later he extended the rails a further 2 kilometres into the village of Llanbedrog.[41] Along much of its length the rails, 3 feet in gauge, had been laid in a vulnerable location right along the back of the shingle beach of Traeth Crugan. Not surprisingly, in 1896, the only recently opened line was damaged severely in a storm. As a result, its rails had to be re-laid further inland. However, the sea was to win ultimately; a very severe storm developed over the Irish Sea on 28 October 1927 and high seas coupled with strong onshore winds affected much of northern Cardigan Bay.[42] There was extensive flooding in Pwlhelli and, yet again, very substantial sections of the line were swept away by breaking waves.[43] Following this devastation, the tramway was not repaired; it was not deemed economic to do so on this occasion and, thereafter, the horses were put to grass.

Beneath the waters of the magnificent sweep of Cardigan Bay lies the fabled Cantref y Gwaelod, or the Lowland Hundred.[44] The submergence of this land – the Welsh Atlantis – is attributed in legend to a drunken gate master called Sicthenym or Seithenyn who, in the sixth century it is said, neglected his duties and left open a water gate that protected it.[45] The legend holds that this was a large fertile area of pastureland that contained sixteen flourishing settlements and was traversed by several roads.[46] It extended from Ramsey Island, off the coast of the St David's peninsula in Pembrokeshire, north to Bardsey Island, off the Llŷn Peninsula. The main sea wall or dyke that protected these lands and their people was known as Sarn Badrig or St Patrick's

Causeway. This is, in fact, a long, natural bank of glacial deposits that extends for over 20 kilometres south westwards from Shell Island near Harlech, parallel with the Llŷn Peninsula. Exposed at low water on spring tides, it forms a hazard to shipping. Similarly the roads, visible at very low tides as causeways, are made of coarse glacial deposits. The demise of the Lowland Hundred was, in fact, due to rising relative sea levels that have progressively inundated this low-lying coastal plain following the Pleistocene glaciations; a rather more prosaic cause than the careless folly of an inebriated gate master but with the same consequences.

Opened in 1867,[47] the Cambrian Railway from Machynlleth to Pwllheli is, like so many others, a major feat of civil engineering. Crossing narrow rivers and wide estuary mouths alike, in the substantial sections where it clings to the coast most closely, the line encounters beach shingle ridges; it intersects salt marshes and estuarine inter-tidal sand and mudflats. In others, it is supported on ledges cut along curves around rocky headlands and slices through cuttings. Elsewhere, it acts as a coastal defence, its embankment forming a masonry sea wall that deprives the beach below of its natural supply of sediment eroded from the adjacent cliffs. In short, along its 90-kilometre route this railway is a microcosm of the whole gamut of terrain types that the Victorian railway engineer had to contend with and, what is more, to overcome. The metals cross the salt marshes and dunes of the Dovey Estuary, curve along a ledge above the rocky shore of the Dovey east of Penhelig, pass through tunnels at Penhelig and Aberdovey, and cling vulnerably to the low-lying coastal edge along the shore north of Tywyn. Carrying the single-track line, another narrow rocky ledge is cut into the slopes of the cliffs between Llwyngwril and Fairbourne followed by more, once open, marshes traversed between Fairbourne and Morfa Mawddach. At Llanaber, heavily protected with rock armour, the railway is again constructed right at the coastal edge, protecting the base of the cliffs and cutting off the natural supply of shingle from the cliffs of glacial deposits to the beach below. Similarly, along the coast from Llandanwg north to Harlech, the railway abuts the foot of the cliffs of glacial deposits, which are now protected by the railway embankment with a vertical masonry wall between it and the boulder-strewn beach below. The famous coastal geomorphologist Alfred Steers pondered as long ago as the late 1930s that, 'It would be interesting to speculate what would have happened if the railway embankment had not been built along the foot of Harlech cliffs, and had not cut off the supply of available shingle.'[48] To the north of Harlech, at Talsarnau, the line again crosses once open marshland, now drained and embanked from the sea. Between Minffordd and Porthmadog it traverses the once broad estuary of Traeth Mawr, claimed from the sea by William Madocks in 1811, on an embankment built to the north of Madocks's great structure, the 'Cob' (see Chapter 4). Another stretch of once open salt marsh is sliced through between Criccieth and Pwllheli, along with storm beach shingle ridges.

Along this railway, man and nature are intertwined at the coastline and there has been a continual battle between it and the sea. Less than two years after it opened it was breached by storm waves in February 1869 and flooded in several localities[49] and again, especially near Tywyn, in a gale in October 1896;[50] a turn of events that was to continue throughout the remaining years of the nineteenth century.[51] Through time and with frequent repairs, the line's defences became strengthened but it has always remained vulnerable to the sea. For instance, a severe gale hit Cardigan Bay on 23 November 1938 and caused great damage along the line; it was severed in several places between Penrhyndeudraeth and Barmouth. The sea wall was breached at Talsarnau and there was extensive flooding of the marshes. All of the low-lying area at Fairbourne was flooded. Nearby, at Llandecwyn, the rails were undermined as the ballast was excavated by the waves. In other places shingle was pushed onshore in the storm and covered the rails. A bridge near the shingle coast between Criccieth and Porthmadog was washed away half an hour after a train carrying about sixty passengers had passed over it. A passenger train that had left Porthmadog for Criccieth was stopped within sight of the wreckage, thus a potential disaster was averted.[52] It would appear that the battle along this coast is again swinging in favour of nature. The storms that heralded in the New Year of 2014 caused extensive devastation to the line; at Tywyn, where the tracks were left hanging up to 60 metres up in the air, at Barmouth and Criccieth and at Llanaber, where a large section of the sea wall was completely washed away. Elsewhere the rails were buried by storm-tossed debris. It is likely that the route will have to be closed north of Harlech for about four months until all of the repairs are completed.[53] By good fortune, no lives were lost along the line as a result of these events. In past times, however, providence has not always been so benign.

A decade before the opening of Solomon's tramway, the New Year of 1883 was to dawn with a terrible tragedy on the edge of this part of

Wales. Readers in the north east of England were amongst the first to learn of the accident when they opened their morning newspapers on the morning of Tuesday, 2 January. The four-part headline in the *North-Eastern Daily Gazette* encapsulated what had happened the night before: 'Terrible Railway Accident in Wales; An Engine over a Precipice; The Engine Driver and Stoker Killed; Marvellous Escape of the Train.'[54] The scene of the accident was close to Friog on the line between Llwyngwril and Fairbourne near Barmouth, the line at this point running along a ledge-like cutting some 15–20 metres above the shores of Cardigan Bay. An engine and tender plunged over the cliff, killing the driver and fireman whose bodies were horribly mutilated by the jagged rocks beneath. The 5.30 p.m. service from Machynlleth to Pwllheli was advancing at its ordinary speed – which, owing to the gradient down to Barmouth, was regulated to a sedate 4 miles per hour – when the engine, *Pegasus*,[55] collided with an obstruction on the line, whereupon the engine and tender rolled over the edge and fell down onto the beach below which is strewn with boulders. Fortunately the other four carriages and van which made up the train did not follow. The first carriage turned over on its side and came to rest partially overhanging the cliff, the coupling between it and the tender having providentially broken.[56] Although several reports noted that it was fortunate that this Third Class accommodation contained no passengers,[57] there was, in fact, one old lady in this overhanging carriage. Mrs Lloyd of Welshpool was a well-known character at Tywyn but apart from severe fright she was freed from the carriage without sustaining further harm.[58] The second carriage of First Class accommodation also turned onto its side amidst the debris, whilst the remaining two did not leave the rails. Apparently the train had been derailed by the debris of a landslip that had fallen onto the line. The retaining wall of a turnpike road which traversed the cliff on a second ledge at a higher level than that of the railway (today this is the A493 road) had apparently given way, bringing rocks and assorted debris onto the metals.[59] Only a few passengers were on the train, about six in total, including the then Vice-Chairman of the Cambrian Railways, Captain R. D. Pryce, who, as might have been anticipated, was travelling First Class. They were all able to escape through the carriage windows practically unhurt. Ironically this was not the first brush with death that Captain Pryce had faced on one of his own company's services. The winter storms of 1868 had caused the River Severn and its tributaries to flood with alarming rapidity and had washed away a section of railway embankment close to Caersws between Machynlleth and Newtown in mid-Wales. In the dim dawn of February an oncoming early morning goods train had tumbled into the void, killing the driver and fireman. En route to the scene on the first passenger train available from Aberystwyth was Captain Pryce, accompanied by the company's traffic manager, when it was discovered that a bridge near Pontdolgoch ('The Bridge over the Red Meadow') was giving way under the force of the torrent. The train, crowded with passengers, was brought to a halt just in time to avert what could have been a major catastrophe.[60]

It was believed that the extent of the Friog disaster was lessened by a second landslip, which took place just as the train was passing. This actually prevented the carriages from following the engine and tender by partially burying them. As this was an area of frequent landslips and rock falls, a watchman was always stationed close to where this accident took place to signal any fall of debris. Apparently he had walked the line prior to the tragedy and found it to be clear. It was, however, conjectured that the vibration caused by the train, along with the recent heavy rains, had initiated the landslide.[61] The jury at the inquest returned a verdict of 'accidental death' but considered the so-called Friog cutting unsafe and hoped that the Cambrian Railway Company would take steps to prevent the recurrence of a similar accident. Captain Pryce said that the company would carry out whatever was recommended by the Board of Trade.[62] This would not, however, be the only occasion when vibration and the loosening of earth materials caused by the passage of a train was to be implicated in a coastal railway accident on the edge of Britain.

In 1922, in his reflective *History of the Cambrian*, the writer and newspaper editor Charles Gasquoine declared rather boldly that: 'Such occurrences, alas! [sic] are not entirely within the compass of human power to control, but, as a matter of fact, no such "similar accident" has during its history ever happened at Friog or anywhere else on the Cambrian system.'[63] Famous last words indeed and, despite Captain Pryce's assurances at the time of the Friog tragedy, half a century later an almost carbon-copy accident occurred in 1933, a short distance to the north of the location of the 1883 event. At a little after 7.00 a.m. on 4 March, a train heading from Machynlleth to Barmouth was struck by a rock fall and hurled over the cliff to the beach below. The driver and fireman were both killed. Fortunately, yet again, the engine coupling broke so that the carriages remained on the line.[64] On this occasion there was only one passenger on the service and he escaped without injury. He gave the following account of the tragedy:

Between Llwyngwril and Fairbourne the train came to a sudden standstill. I was sitting in the rear coach near the guard's van, when there was a terrible vibration like an earthquake. My breath seemed to have stopped temporarily. It was a stunning sensation. The guard and I jumped down and saw the landslide. The engine had been uncoupled by the strain and had fallen over the rocks on to the seashore.[65]

Again the jury at the inquest returned a verdict of 'accidental death' and expressed the opinion that a permanent watchman should be reinstated on this section of the line, a practice that had clearly lapsed in the interim, and that the speed limit, which had been increased since the 1883 event, should be reduced from 15 to 10 miles per hour.[66] Today the line at Friog passes through a sturdy, rock avalanche shelter. However, it has to be reiterated that the various measures that have been taken to stabilise the cliff along the length of this cutting face now reduce the natural supply of rock fall material to the beach that there would have been prior to the arrival of the railway.

The rocks of this part of Wales are, perhaps unsurprisingly, of Cambrian age, some 500,000,000–540,000,000 years old. In the Fairbourne to Llwyngwril district the sequences comprise typically grey, finely laminated to thinly bedded siltstones, mudstones and shales that have been altered through metamorphism and by deformation through time.[67] Indeed, the layered and deformed character of the rocks in the Friog cutting is clearly captured in a contemporary painting of the 1883 accident (Figure 2.5). These materials are cross cut by several sets of fractures – joints, faults and planes of cleavage – that render the sequences as unstable or potentially unstable at cutting slopes. Furthermore the fractures in these rocks – collectively known as discontinuities – are closely spaced and are inter-connected, which is favourable for the flow and migration of groundwater through the rocks and a major factor influencing rock slope stability.[68] In his official report into the 1933 slope failure, Lieutenant Colonel A. H. L. Mount, the Chief Inspecting Officer of Railways, drew attention to the geology of the rocks in the Friog area and reiterated comments made by his predecessor in the role, Colonel F. H. Rich, at the time of the 1883 accident. Colonel Rich was bluntly forthright about the geology of the site. He

Figure 2.5 The accident at Friog cutting, 1 January 1883. Note the inclined and finely laminated 'bastard shaly rock' in the background along with the turnpike road on a second ledge at a higher level above the railway (image supplied courtesy of Llyfrgell Genedlaethol Cymru/The National Library of Wales).

described the headland or edge of the mountain slope, around which the railway and road run, as consisting of:

> . . . a bastard shaly rock which disintegrates into small pieces on the surface. A great part of it is over-laid with a peaty loam, which is full of stones and grows heather and grass. The thickness of the loam overlying the rock varies very much.[69]

Moreover, as another reminder of the dynamism of this part of the coast, a landslide at Friog closed this railway line yet again in late March 2005.[70] The slip was caused when the seawall became breached by storm waves during a high tide, causing stone blocks and debris to be pulled down behind it from the cliff and onto the track bed.[71] The collapse of the roof of a sea-eroded cave below was also implicated in what ultimately led to a fortnight's closure of the line and a lengthy detour by bus for travellers between Machynlleth and Pwllheli.[72]

Boring Barmouth Bridge

As if the Cambrian coast railway did not have enough to contend with its seemingly incessant wave attack and landslides deflecting its locomotives off the metals and into the sea, an insidious foe was also at work. The small resort of Barmouth is located at the seaward end of the north shore of the estuary of the River Mawddach. A wooden bridge that carries the single line railway – the longest such structure on the British railway system – spans the mouth and is a critical link, reducing the length of the journey by road around the head of the estuary by over 20 kilometres. Barmouth Bridge (Figure 2.6) was originally built in 1866–7 and is 820 metres in length. Reconstructed in 1903–6, it has a timber superstructure for most of its length, which is supported by 113 groups of timber piles.[73] In April 1980, during routine maintenance, it was realised that the bridge had succumbed to a serious attack of shipworm. This mollusc, the organism *Teredo sp.*, bores surreptitiously into wood and rock creating holes up to 2 centimetres in diameter. The bulk of the deterioration was at bed level, where pools of water trapped around the bases of the pillars encouraged the infestation.[74] This initially necessitated a weight restriction on the bridge, so causing the suspension of goods traffic to the north of it. By October 1980,

Figure 2.6 View to the south west over the Mawddach Estuary and Barmouth Bridge towards Fairbourne. The Cambrian coast railway to Fairbourne is located at the foot of the hills in the distance. The Fairbourne Steam Railway extends from the village along the coastal edge of the length of the curving spit in the foreground to Barmouth Ferry Station at its extremity (photo courtesy of Wyn Edwards).

however, the full extent of the damage was determined; sixty-nine of the pillars had been eaten into and, as a result, the viaduct was closed to all trains. This threatened the whole route's survival. Temporary upkeep permitted the bridge to be reopened the following May, but for light trains only. The complete and extremely costly repairs were made gradually, being completed in 1986 when the bridge was re-opened ceremoniously to locomotive-hauled trains. Most of the piles, 330 out of a total of 500, were encased in glass-reinforced concrete jackets with resin injected under pressure between the jackets and the timber to kill off the shipworms. Forty-eight of the piles were beyond repair and replaced completely with like-for-like Greenheart hardwood.[75]

Somewhat ironically, it was an almost-local Welsh geologist from Montgomeryshire who, in 1918, first introduced the non-specialist, non-zoologist to the impact that boring organisms can have at the coast, eroding rock and man-made structures.[76] Dr Thomas Jehu, the first Lecturer in Geology at Queen's College, Dundee, appointed in 1903, was a member of the *Royal Commission on Coastal Erosion* that reported in 1911. That same year he described the glacial deposits of western Caernarvonshire, drawing attention to their serious erosion in the cliffs close to the Cambrian Railway between Pwllheli and Criccieth.[77] Subsequently, in 1914, Jehu was to become the Regius Professor of Geology and Mineralogy at the University of Edinburgh. It is unlikely that even he could have foreseen the extent of the damage that shipworms would ultimately inflict upon the Barmouth Bridge.

Across the sea

Lest it be thought that the toppling of trains into the sea owing to collision with a landslide was a British peculiarity, across the sea in Northern Ireland a parallel disaster took place on the shores of Belfast Lough where the track is right at the water's edge. It was on the winter's evening of Saturday, 19 February 1910 that a mail express en route from Belfast to Larne smashed into displaced debris on the stretch between Carrickfergus and Whitehead. Hundreds of tonnes of sand and clay had become dislodged by heavy rains associated with a severe gale and had completely buried the track at a place called Brigg's Loop.[78] The accident took place along a narrow coastal strip where the rocks are principally inter-bedded soft, friable mudstones and sandstones of Triassic age. These are prone to instability, slippage and marine erosion, resulting in the construction of numerous coastal defences over the years. The driver of the advancing train had no warning to apply the brakes and it bulldozed into the obstruction; the engine and first three carriages forcing their way into the debris at a speed of 40–50 miles per hour. The next carriage, however, was lifted up from the rails and it slid down the slope of the embankment into the sea, dragging the succeeding two coaches with it. About forty passengers were in the derailed coaches, two of which were completely overturned, their roofs coming to rest on the shingle at the shore.[79] Although many passengers were flung onto their heads and water surged into the compartments, only three were seriously hurt though several sustained injuries. The water was around 3–4 feet in depth at the time of the derailment. The following Saturday, 26 February 1910, the *Penny Illustrated Paper* published a grainy photograph of the disaster scene along with another showing the shocked and bewildered passengers walking along the waterside track, bestrewn with wreckage, transferring to another train that had been despatched to rescue them.[80]

The sea breaks through

The water rushed over the road and into the adjoining fields with great force, and so strong was the tidal flood that a breach was made in the Lytham and Blackpool Coast Line, at South Shore, behind the Manchester Hotel. This breach is about fifteen or twenty yards wide, and runs right through an embankment leaving nothing but the rails and sleepers hanging crookedly, in skeleton shape, from one side to the other.

Great storm at Blackpool: Roads torn up – fields
flooded – a railway embankment broken through.
The Preston Guardian, 21 September 1872

A perennially vulnerable railway circumnavigates the Cumbrian coast. Opened in stages between 1844 and 1866,[81] the line extends from Carlisle to Barrow-in-Furness. For most of the 180-kilometre-long route it hugs the coast, but in so-doing traverses widely differing terrain. At Whitehaven, for instance, the line perches on a sea wall above a boulder-strewn rocky platform, whereas a little to the north, at Parton, it is located at the back of a beach through low sand dunes at the foot of landslide-prone cliffs. Elsewhere, at the coast, the line is built along several substantial sections of sea wall, over shingle ridges and dunes. The Parton to Harrington section is especially prone to damage by the sea, as is the line at Flimby near Maryport and between St Bees and the remote station at Nethertown. The Cumbrian coast was particularly badly hit by the series of tremendous westerly gales coupled with heavy falls of rain that brought 1852 to a close. Numerous vessels were driven onto the shore and wrecked.[82] The railway at Whitehaven was very seriously damaged close to the town, a repeat of events of earlier that same year:

> The damage caused by the storm of last winter, and the locality of it, will be fresh in the minds of our Whitehaven readers. The earliest damage inflicted on the line by the storm of yesterday was the destruction of a portion of the old sea wall, immediately to the north of the new one which had been erected towards the repair of the disaster of last February.[83]

The force of the breaking waves was such that, having once made a breach in the wall, they tore away the foundations of the railway behind it and ripped out great sections – hundreds of metres in length – of the rails along with the wall on the opposite side of the tracks.[84] 'A tremendous breach was thus created by the removal of the whole of the new wall and a large extent of the old one adjoining.'[85] Extensive damage to the line occurred further to the north. At Risehow, near Flimby, for instance, nearly a mile of the rails was washed out entirely. At that locality the sleepers had been laid directly on the dunes at the back of the beach – there was no wall to protect the line.

> Altogether the railway presents a sad and melancholy wreck. Heavy stones have been thrown upon the line within reach of the tide, telegraph posts blown down and the wires broken, the ballasting of the line washed away in various places, and in many respects it exhibits unmistakeable proofs of the fury of the tempest.[86]

In the February 1869 gale that damaged the Cambrian coast line, the rails at both Flimby and St Bees were again torn up and there was considerable damage to the wall carrying the line at Whitehaven and to the wall that protects it at Harrington.[87]

Breaches by the sea leading to continual repairs and the periodic installation of additional protection measures have been a constant feature of this line throughout the twentieth century,[88] right up to the present day (Figure 2.7). The January 2014 storm, as a recent repeat of

Figure 2.7 A nuclear flask train en route in July 2012 to the spent nuclear fuel processing facility at Sellafield. The train is passing along a section of the Cumbrian coast line at Nethertown newly repaired with rock armour (photo courtesy of N. Booth).

so many previous episodes, removed almost 200 metres of the track at Flimby, one of the 'weak links' along the route.[89] Moreover, the patchwork of the wall north of Whitehaven bears witness to the numerous attempts that have had to be made over the years to maintain its integrity (Figure 2.8).

Elsewhere in Britain, some of the early standards of railway engineering were similarly no match for the sea.[90] Even before it had carried a 'Directors' Special' on 17 September 1852, the South Wales Railway between Swansea and Carmarthen was breached by the sea.[91] This extension of the route westwards via Llanelli involved a particularly vulnerable section along the northern shore of the Loughor Estuary between Llanelli and Burry Port known as Cefn Padrig. It was here that the initial damage by the sea took place, where the line was said to make 'a long flat run along the sands.'[92] Flat it is, but most of this bay is now muddy. The damage was made good and the lengthy special train was able to make its passage safely. The following month the route opened to the general public but again, just over two months later in December 1852,

Figure 2.8 Looking northwards along the Cumbrian coast railway at Tanyard Bay to the north of Whitehaven. The masonry sea wall that carries the line has been strengthened by rock armour and concrete (photo: R. W. Duck).

the same storm that caused so much damage at Whitehaven ruptured the low embankment.[93] It was once more repaired and the protective sea wall strengthened but not sufficiently so to withstand a gale in January 1890[94] and, even worse, in a violent storm a little over six years later it succumbed to the waves again with the resulting damage amounting to thousands of pounds. On 8 October 1896, during the same storm that damaged Pwlheli and the Cambrian coast line, the morning mail from London had passed through Llanelli without incident. However, almost as soon as it had had done so, the massive stonework of the sea wall gave way.[95] 'The permanent way was next torn up, and a roadway 150 ft. wide by ten feet deep was cut under the metals; the sleepers remained but the metals presented the appearance of a switchback railway.'[96] Not all of our coastal railways were, however, to experience such a fortunate escape from fatal disaster during their early years.

The *Irish Mail* was the first named train in the history of Britain's railways and it made its inaugural journey in 1848 with the opening of the Chester and Holyhead Railway.[97] The construction of the 136-kilometre route along the North Wales coast, across the Menai Strait to Anglesey and onward to Holy Island, was yet another major civil engineering achievement; more so because Parliament had only agreed to its construction four years previously. The line was to become particularly famous for its two tubular wrought iron bridges designed by Robert Stephenson; the Britannia Bridge across the Menai Strait and the Conwy Bridge across the tidal mouth of the river of that name. But, as the Resident Engineer for these unusual structures, Edwin Clark, remarked of the line: 'A series of works of unrivalled magnitude characterises its whole length of 84½ miles.'[98] The line crosses marshlands and skirts estuarine shores; it is tunnelled through several rocky headlands but for much of its route it is right at the coastal edge on top of a series of sea walls. The line's 'sea walls and defences', noted Clark in 1850,

> . . . on the one hand, along this exposed coast, are all on a magnificent scale; whilst, on the other, a timber gallery, similar to avalanche galleries on the Alpine roads, protects the road line from the *débris* that rolls down from the lofty and almost overhanging precipices above it.[99]

The link between Anglesey and Holy Island was made via the 1,200 metres long Stanley Embankment, a pre-existing causeway, built across Beddmanach Bay in 1823 and designed by Thomas Telford, that was widened to accommodate the railway.[100] The structure is named after Sir John Stanley of Penrhos, near Holyhead, who helped to finance it and whose family played a major role in facilitating the coming of the railway to the island and in developing Holyhead as a port.[101] Like so many railways in coastal settings, it was not long after its opening that the sea cast the first stone and began to take advantage of its most vulnerable sections. The widened Stanley Embankment was one such weak link and although the sea walls on the mainland might have been built on a 'magnificent scale', they were no match for the power of the waves at several exposed localities. Vulnerable sections were at Bagillt on the Dee Estuary coast to the north west of Flint and along the more open coast of Rhyl further to the west. These became damaged or washed away on several occasions, most notably on 26 October 1859 and thereafter,[102] in the same storm that sank the Australian emigrant ship, *Royal Charter*, with the loss of over 400 lives.[103] The Stanley Embankment was very badly damaged in a gale and partly washed away by the sea in November 1887; elsewhere along the line there were several breaches.[104]

Engines had crossed the link with Holy Island with caution as the water level rose until eventually, 'it went in with a crash and the waves made clean breaches over the line and ran right up to Valley Station, some two miles distant.'[105] Similarly, at Rhyl, the railway was swept away in a storm in October 1896 in which 'immense blocks of granite, weighing from 15 cwt. to two tons were lifted up like paper, and flung bodily over the embankment'.[106] Decades later in 1945 and almost a century on in 1990, the embankment at Towyn (not to be confused with Tywyn on the Cardigan Bay coast), between Abergele and Rhyl, was yet again breached by storm wave attack.[107] And on many other occasions, overtopping by waves between Rhyl and Abergele during gales and the consequent inundation has caused this line to be closed to traffic.[108]

At about an hour before midnight on Thursday 12 January 1899 a much more serious disaster took place approximately mid-way between Conwy and Penmaenmawr. It was said that: 'The night was pitch dark and the tempest raged fiercely' and the tide owing to the terrific south-westerly gale that was in progress at the time had risen 'much above its normal level'.[109] As a result of the incessant wave attack on the railway embankment wall, the structure eventually gave way at the western end of the Penmaenbach tunnel and the water inundated the railway line. The breaking waves and ensuing backwash thus removed the ballast from beneath both the up and the down rails for a length of some 70 yards, leaving the rails and sleepers suspended in mid-air without any foundation. Just at this time, around 11 p.m., an express goods train en route from Manchester to Holyhead was speeding westwards in the darkness and made direct for the gap. The weight of the engine and its tender immediately brought down the unsupported rails and both vehicles toppled headlong into the stormy sea. It is thought that the driver and the fireman must have been drowned almost immediately. There were thirteen trucks and vans making up the train, loaded with various goods. Eight followed the ill-fated engine, but five stopped short of where the line had been undermined by the sea.[110]

The engine was found when the tide had receded, lying on the pebbly beach flat on its side with a length of rail curling under it. It was par-tially covered with debris, the tender having crashed down on top of it. The trucks lay in a confused heap of wreckage. The third member of the train's crew, the guard in charge of the train and a brakeman who travelled with him, had a merciful escape as they had been in the van at the rear of the train at the time of the tragedy. The disaster would have surely resulted in far greater loss of life had the unfortunate train been carrying passengers. It was thanks to the brakeman in raising the alarm that an up Liverpool express was brought to a standstill, 'just before it reached the brink of the peril'[111] and its passengers were taken safely to Conwy. In order to make swift repairs, a new set of rails was laid on the landward side of the former rails with the necessary ballast being obtained from local quarries.

At the inquest into the disaster the guard described that he had seen a heavy sea breaking right over the train. A foreman platelayer gave evidence as to the steps taken on this and other stormy nights to watch the tide and protect the line. He had shouted and waved his lamp in an attempt to attract the driver's attention, but the elements made such uproar that his voice could not be heard and the subsequent accident could not be averted. The jury returned a verdict of 'accidental death' and recommended that a signal box be built, to be open day and night, at the western end of Penmaenbach Tunnel. Moreover, they advised the construction of a more substantial sea wall to protect the railway.[112] Construction on a merely 'magnificent scale' was inadequate for this length of coast. The response to the damage was swift: 'The railway company's officials were engaged yesterday in lowering immense granite slabs into the gap to constitute a stronger foundation for the new sea wall.'[113] However, in the immediate aftermath of the disaster, it was left to *The Times* to bring another, typically Victorian, calming sense of perspective to the incident, by reporting that: 'One or two of the wagons which fell into the chasm contained barrels of stout, and these were found intact'.[114]

The Line to Legend Land

In the early 1920s, the Great Western Railway marketed its network as 'The Line to Legend Land'. This was a reference to a whole host of old tales and folklore that could be used to potentially lure would-be tourists and other visitors to its numerous destinations. At the coast, where reality and myth are never far apart, the lost land of Lyonesse was right at the heart of one such magnetic legend:

Beneath Land's End and Scilly rocks
Sunk lies a town that Ocean mocks

Even the railway company's tales were purported to be retold by 'Lyonesse'.[115] Three British steam locomotives bore the name *Lyonesse*, the first of which, and obviously a Great Western, was in operation from the beginning of the twentieth century until the 1930s.[116]

This fertile 'land of matchless grace', famed in medieval Arthurian legend as well as Cornish and Breton Celtic mythology, once existed offshore from Penzance, between Land's End and the Scilly Islands. The latter group, along with the small tidal island of St Michael's Mount, are all that remain of this lost, sunken land which, like Cantref y Gwaelod, was a victim of rising sea levels.[117] It is perhaps ironic that a legend pertaining to the loss of land to an encroaching sea should be used by the Great Western Railway to encourage its passengers to visit the utmost limits of its network in Cornwall and Devon when the line itself traverses, thanks to massive feats of engineering, some dangerous and vulnerable shores. In this regard, the Devon resort of Dawlish, en route to the Penzance terminus, is without equal:

> Reaching down where the fresh and the salt water meet,
> The roofs may be seen of an old-fashioned street;
> Half village – half town, it is – pleasant but smallish,
> And known, where it happens to *be* known, as Dawlish.[118]

At Dawlish the railway is the sea wall and the sea wall is the railway. Some authors have not hidden their abhorrence of this structure, interposing, as it does, a barrier between town and sea: 'Even amid the excesses of the twentieth century', it has been written, 'this example of Victorian vandalism is unforgiveable.'[119] Sir John Betjeman, however, painted a romantic picture of springtime idyll:

> Bird-watching colonels on the old sea wall,
> Down here at Dawlish where the slow trains crawl:
> Low tide lifting, on a shingle shore,
> Long-sunk islands from the sea once more:
> Red cliffs rising where the wet sands run,
> Gulls reflecting in the sharp spring sun;
> Pink-washed plaster by a sheltered patch,
> Ilex shadows upon velvet thatch:
> What interiors those names suggest!
> Queen of lodgings in the warm south-west. . .

But nothing could be further from the truth in winter months when the sea wall takes the full force of the gales. This impressive structure, engineered by the great Brunel has been put to the test on many occasions since it was built. Furthermore, it has lost the battle with the sea on equally as many occasions:

It has come at last! The conjunction of south-east gales, with spring tides, has completely put a stop to the traffic on the South Devon Railway. Those who in their rambles have marked the inroads which for years past the sea has been making at certain points along the line, from Cockwood to Teignmouth, felt convinced that the time must come when railway works would succumb to the incessant beatings of the tide. More than once already the traffic has been interrupted and delayed by accidents occasioned by heavy seas, at Langstone Point, the sea wall at Teignmouth, and one or two other places. These, however, were minor affairs compared with the calamity which befell the line last evening, when three breaches were made between Starcross and Dawlish and all traffic was suspended.[120]

This might well have been a description of the dramatic events – and made so in no small part thanks to widespread media coverage – of the storms of January and February 2014. It is, however, from over a century and a half earlier in 1859. On the night of 4 February 2014 and in the ensuing gales, the wall was breached along a length of about 40 metres and the rails left hanging in mid-air as breaking storm waves ferociously dug out the ballast and fill from beneath. The force of the waves was such that houses to landward of the line were also damaged. Elsewhere the wall and track bed was damaged extensively, fences were demolished; and ballast was washed out, tossed in the air by waves and left strewn on nearby roads. Battered and pounded by waves for several days, the wall and rails were left covered with debris. The last surviving rail artery to south-west England had been completely severed (Figures 2.9 and 2.10).

The Dawlish sea wall, built in 1845 and opened to traffic the following year, initially carried Brunel's broad gauge (7 feet, 4¼ inches), single track coastal line of the South Devon Railway. Essentially Brunel had two options; he could either build along the shoreline or via lengthy tunnels between Dawlish and Teignmouth (Figure 2.11). He chose the former, which necessitated the construction of five relatively short tunnels, through protruding rocky headlands, along with about 6.5 kilometres of sea wall. Heading south from Dawlish the tunnels

Figure 2.9 The breach in the South Devon Sea Wall at Dawlish photographed on 7 February 2014 after the rails had been cut away; the cut sections with sleepers attached are lying in the gap. The Permian age, red sandstone cliffs are conspicuous in the background ('The breach in the South Devon Railway sea wall at Dawlish, Devon, England, February 2014'; from Wikipedia by Smalljim, licensed under CC BY-SA 3.0).

Figure 2.10 The aftermath of the storm along Marine Parade and the adjacent railway, looking towards Dawlish station, on 8 February 2014 (photo: R. W. Duck).

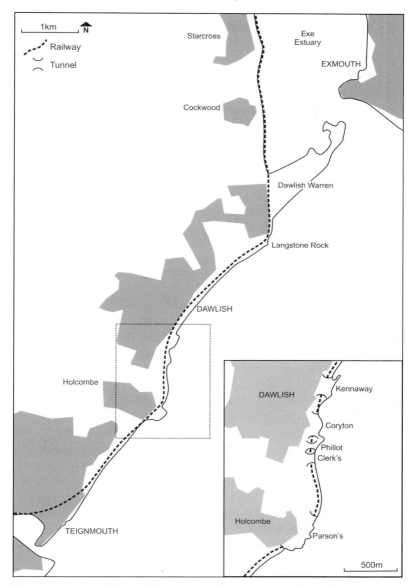

Figure 2.11 Map of the Dawlish–Teignmouth area showing the South Devon railway line and localities referred to in the text. The inset shows the five tunnels along this part of the line (drawing by T. Dixon, based on © Crown Copyright/ database right 2013. An Ordnance Survey/EDINA supplied service).

are; Kennaway (187 metres in length), Coryton (207 metres), Phillot (45 metres), Clerk's (53 metres) and Parson's (342 metres). At Dawlish the formation directly hugged the foot of the cliff; elsewhere – as near Dawlish Warren – the embankment was originally built along the shore isolating swampy ground between it and the cliff. Nearby a deep and narrow cutting took the line through Langstone Point but, through time as the line became widened, this promontory became isolated to form Langstone Rock. In sum, it was a masterpiece of coastal railway engineering. First of all, the South Devon Railway's trains were hauled by steam locomotives but in 1847 became propelled by the atmospheric system, being forced forwards by atmospheric pressure through a pipe. Hailed as silent, smokeless and swift, after only eight months of use, the system was abandoned owing to its high costs and to its unreliability in both hot and icy weather conditions. Its lack of success is widely regarded as Brunel's greatest failure. In 1892 the line was converted to standard gauge and by 1905 it had been completely doubled through the tunnels, involving widening them, and along the sea wall (Figure 2.12).[121]

Figure 2.12 The Dawlish sea wall and railway looking from the north portal of Kennaway Tunnel along Marine Parade (photo: R. W. Duck).

Right from the very start of construction, the sea wall proved to be problematic for Brunel; the eternal battle with the sea had commenced. A gale in October 1846 claimed no less than eight breaches in the structure between Clerk's Tunnel and Parson's Tunnel; Brunel himself went to inspect the damage and plan how best to make it good.[122] Shortly afterwards, on 21 November 1846, *The Times* reported that:

The two aquatic heroes – Neptune and Brunel – who alone divide public interest in this part of the world, have had another severe bout, which has ended in the engineer being thrown a tremendous back fall, which took him so completely aback that he will not soon recover from its effects.

On the previous evening, a southerly gale had coincided with a high spring tide. The violence of the storm:

proved disastrous to the exposed surface and embankments of the South Devon line, which had been scarcely set to rights from the damage sustained by the gale about a month since, which did such immense mischief to the coasts of Devon and Dorset.[123]

In this event, the second in close succession to have had an impact, yet again much damage and several breaches of the sea wall occurred between Starcross and Dawlish. Indeed, over the subsequent decades in this area, the combination of southerly or south-easterly gales and high spring tides during the winter months was to frequently prove the most powerful and damaging weapon in Neptune's armoury.

On 17 February 1855, the wall once again gave way to wave attack, on this occasion at the foot of Smugglers' Lane, Holcombe, close to the southern portal of Parson's Tunnel. It was reported that upwards of 175 passengers on a train from Teignmouth were exposed, for some considerable time, to the rigours of the gale. They were obliged to clamber – fortunately without injury – around the breach to join a train from Exeter that had been sent to meet them,[124] the whole experience being visualised romantically by *The Illustrated London News* a fortnight later (Figure 2.13). Other major breaches of the wall and railway at Dawlish took place in February 1869[125] and on Hogmanay 1872,[126] the former the result of the same storms that devastated both the Cambrian and Cumbrian coast lines. And so the story goes on. Major rebuilding, enlarging and strengthening of the structure along

Figure 2.13 The breach in the Dawlish sea wall close to the southern portal of Parson's Tunnel. Note the stack of Parson Rock, after which the tunnel is named, protruding from beneath the sea beyond the breach (from *The Illustrated London News*, 3 March 1855).

the Dawlish waterfront of Marine Parade began in 1901. This caused substantial draw down of the beach level at its foot with the result that waves of even greater height could break directly against the wall and even over passing trains; today the lack of sand and shingle on the beach along the Dawlish waterfront is readily apparent. The wall was breached again by the sea in various places, amongst other occasions, in January 1930,[127] in December 1945[128] and in February 1986[129] and in a storm in February 1974, waves 20 metres in height broke over Dawlish station, demolishing the down platform and leaving the wreckage strewn across the tracks.[130]

The Dawlish sea wall has had a major impact, altering the natural coastal system in the area in terms of the physical processes operating. Whilst the wall has effectively cut off the sediment supply from the naturally eroding cliffs to the beach and along the shore, the cliffs are still susceptible to erosion by the largest of storm waves as these have the capability to overtop the wall and the railway entirely and periodically break onto the rocks to landward. But despite repeated and extensive storm damage, breaches, derailments and constant pounding by the sea, the Dawlish sea wall has never claimed the life of a railway traveller.[131] It was the cliffs through which the line was cut or tunnelled that were to be the cause of both problems and tragedy, as described in Chapter 3. The rocks of this part of the Devon coast are of Permian age – the Dawlish Sandstone Formation – and comprise sedimentary units; principally reddish brown sandstones inter-bedded with coarser, pebbly and even cobble beds known as breccias. The finer sandstones are the product of wind-blown accumulation whereas the coarser beds were laid down by fluvial action. It is a geological paradox that these rocks will stand as near to vertical cliffs yet their component particles are poorly cemented together such that they will readily crumble to the touch between the thumb and the forefinger. Rock falls are, not surprisingly, a frequent occurrence and several sections along the line are netted to prevent this happening or have fences and ditches at their foot to trap fallen debris before it might tumble or bounce onto the tracks, potentially causing a derailment.

To the north east of Dawlish, protruding into the mouth of the Exe Estuary, is Dawlish Warren. This is a 2-kilometre-long spit of sand and shingle, traversed at its landward end by the railway line, which projects across the estuary towards Exmouth. Uniquely, it is, in fact, a double spit. The Inner Warren, the older spit, which is on the northern inland side, is a stable feature about 250 metres across that hosts a links golf course. The younger, Outer Warren on the southern seaward side is topped by active sand dunes but is vulnerable to breaching by the sea and has been eroded severely on many occasions over the past few centuries.[132] In consequence, several attempts have been made to protect it by the installation of a variety of rock armour, gabion baskets and groynes. Between the Inner and the Outer Warrens is a low-lying, formerly tidal creek. Known as Greenland Lake, it is now cut off from the sea and stabilised by vegetation. Warren Point, the furthest projection into the Exe is highly unstable and can change considerably but unpredictably in position and extent according to weather and tidal conditions.[133] Dawlish Warren has been gradually reducing in size since the mid-1800s, when the railway arrived. Although it has been suggested that Dawlish Warren was eroding before the railway was built,[134] there is no doubt that the insertion of the lengthy sea wall between the cliffs and the beach – extending almost continuously from Dawlish Warren to Teignmouth – either cut off or drastically reduced the natural supply of debris to the beach. Furthermore, along with a number of breakwaters, it has helped to starve the natural supply of material to the wave-driven, longshore current transport system that predominantly conveys sediment particles in a north-easterly direction towards Dawlish Warren and across the mouth of the Exe Estuary.[135] The erosion of Dawlish Warren is, as at so many other places around our coasts, one of the additional spin-offs of railway construction that was neither foreseen nor even considered at the time.

Notes

1. Rose, Major C. F. (1981), *Department of Transport Railway Accident Report on the Collision that Occurred on 22nd October 1979 at Invergowrie in the Scottish Region British Railways*, London: Her Majesty's Stationery Office, 22 pp.
   Duck, R. W. (2011), 'The physical development of the Tay Estuary in the twentieth century and its impact', in Tomlinson, J. and Whatley, C. A. (eds), *Jute no More: Transforming Dundee*, 52–69, Dundee: Dundee University Press.
2. Railways Archive, available at http://www.railwaysarchive.co.uk/eventlisting.php?view=chart&showSearch=true& (last accessed 26 April 2014).
3. For example: 'Another railway accident', *Freeman's Journal and Daily Commercial Advertiser*, 16 August 1851.
4. 'The Railway Accident Season', *The Saturday Review*, 8 August 1863.
5. 'Articles', *Aberdeen Weekly Journal*, 2 September 1891.
6. 'Alarming accident on the Highland Railway', *The North-Eastern Daily Gazette*, 27 November 1885.
   'Alarming railway accident', *The Morning Post*, 27 November 1885.
7. 'Extraordinary railway accident', *The Sheffield & Rotherham Independent*, 27 November 1885.
8. 'Extraordinary railway accident', *The Sheffield & Rotherham Independent*, 27 November 1885.
9. Price, M. R. C. (1986), *The Pembroke and Tenby Railway*, Oxford: The Oakwood Press, 112 pp.
10. 'Frightful railway accident', letter to *The Times*, 25 August 1866.
11. Rich, F. H. (1856), Great Western Railway, *Report of The Secretary of the Railway Department, Board of Trade*, 60–61.
12. Rich, F. H. (1856), Great Western Railway.
13. 'A tour to Tenby', *The Leicester Chronicle and the Leicestershire Mercury*, 25 August 1866.
14. 'A tour to Tenby', *The Leicester Chronicle and the Leicestershire Mercury*, 25 August 1866.
15. 'Dreadful railway accident near Granton – four lives lost', *The Scotsman*, 9 July 1860.
16. 'Dreadful railway accident near Granton – four lives lost', *The Scotsman*, 9 July 1860.
    'Dreadful railway accident', *The Times*, 10 July 1860.
17. 'Dreadful railway accident', *The Illustrated London News*, 21 July 1860.

18. 'Fatal railway accident at Ramsgate', *The Morning Post*, 1 September 1891.
19. 'Serious accident at Ramsgate Station', *Birmingham Daily Post*, 1 September 1891.
20. 'The Ramsgate accident', *The Standard*, 10 September 1891.
21. 'To Boulogne and back in one day', *The Times*, 26 June 1843.
22. 'To Boulogne and back in one day', *The Times*, 26 June 1843.
23. Folkestone Harbour Company: History of Folkestone Harbour and Cross Channel Links, available at http://www.folkestoneharbour.com/pages/history. html (last accessed 26 April 2014).
    Hendy, J. (1991), *Folkestone Boulogne 1843–1991*, Kent: Ferry Publications, 40 pp.
24. Hendy, J. (1991), *Folkestone Boulogne 1843–1991*.
25. Sekon, G. A. (1895), *The History of the South-Eastern Railway*, London: Railway Press Co. Ltd, 40 pp.
    Hutchinson, J. N., Bromhead, E. N. and Lupini, J. F. (1980), 'Additional observations on the Folkestone Warren Landslides', *Quarterly Journal of Engineering Geology*, 13, 1–31.
    Hendy, J. (1991), *Folkestone Boulogne 1843–1991*.
26. Hutchinson, J. N., Bromhead, E. N. and Lupini, J. F. (1980), 'Additional observations on the Folkestone Warren Landslides'.
27. Hutchinson, J. N., Bromhead, E. N. and Lupini, J. F. (1980), 'Additional observations on the Folkestone Warren Landslides'.
28. The Remembrance Line, available at http://www.theremembranceline.org.uk/ (last accessed 26 April 2014).
29. Folkestone Triennial, available at http://www.folkestonetriennial.org.uk/2011-event/artists/paloma-varga-weisz/ (last accessed 26 April 2014).
30. 'Railway accident at Folkestone', *The Morning Post*, 15 August 1851.
31. 'Another railway accident', *Freeman's Journal and Daily Commercial Advertiser*, 16 August 1851.
32. 'Alarming accident on the South Eastern Railway', *Lloyd's Weekly Newspaper*, 17 August 1851.
33. Wynne, G. (1851), South Eastern Railway (Appendix No. 64), *Report of the Office of Commissioners of Railways*, Board of Trade, 203–204.
34. 'Alarming accident on the South Eastern Railway', *The Examiner*, 16 August 1851.
35. 'Alarming accident on the South Eastern Railway', *The Examiner*, 16 August 1851.
36. 'Alarming accident on the South Eastern Railway', *The Examiner*, 16 August 1851.
37. 'Pwllheli as a probable watering place', *North Wales Chronicle*, 15 December 1860.
38. 'Pwllheli as a probable watering place', *North Wales Chronicle*, 15 December 1860.
39. 'Pwllheli as a probable watering place', *North Wales Chronicle*, 15 December 1860.
40. BBC Wales (23 May 2011), 'Solomon Andrews, Cardiff Entrepreneur', available at http://www.bbc.co.uk/blogs/wales/posts/solomon_andrews_cardiff_entrepreneur (last accessed 26 April 2014).
    'Pwllheli tramroad', *North Wales Chronicle*, 26 May 1894.

41. 'Pwllheli tramway', *North Wales Chronicle*, 4 August 1894.
Andrews, J. F. (1995), *The Pwllheli and Llanbedrog Tramways*, Cowbridge: D. Brown and Sons Ltd, 72 pp.

42. 'Crane blown over in London', *The Times*, 29 October 1927.
'The gale', *The Times*, 31 October 1927.
'Flood damage', *The Times*, 31 October 1927.

43. Andrews, J. F. (1995), *The Pwllheli and Llanbedrog Tramways*.

44. Ashton, W. (1920), *The Evolution of a Coast-line: Barrow to Aberystwyth and the Isle of Man, with notes on Lost Towns, Submarine Discoveries etc.*, London: Edward Stanford Ltd, 302 pp.
Jones, O. T. (1920), The origin of the Welsh legends, *Welsh Outlook*, 8, 309–312.
Davies, D. J. (1927), Cantref y Gwaelod, *Transactions and Archaeological Record, Cardiganshire Antiquarian Society*, 5, 21–33.

45. Willson, B. (1902), *The Story of Lost England*, London: George Newnes Ltd, 192 pp.
Pennick, N. (1987), *Lost Lands and Sunken Cities*, London: Fortean Times, 176 pp.

46. Willson, B. (1902), *The Story of Lost England*.

47. 'Completion of the Welsh Coast Railway', *North Wales Chronicle*, 19 October 1867.

48. Steers. J. A. (1939), 'Sand and shingle formations in Cardigan Bay', *The Geographical Journal*, 94, 209–227.

49. 'The storm and high tides', *Liverpool Mercury*, 2 February 1869.
'The storm and floods', *The Morning* Post, 3 February 1869.
'The storm', *Birmingham Daily Post*, 2 February 1869.
'Terrific floods in Wales', *North Wales Chronicle*, 13 February 1869.

50. 'Fatal gale and floods', *The Standard*, 9 October 1896.
'Gale & floods', *Liverpool Mercury*, 9 October 1896.

51. 'The gale on the Welsh coast', *North Wales Chronicle*, 3 February 1877.
'The storm and high tides', *The Morning Post*, 20 October 1883.

52. 'Gale and flood', *The Times*, 24 November 1938.

53. BBC News (20 January 2014), 'Cambrian coast rail flood repairs at Barmouth and Pwllheli to take months', available at http://www.bbc.co.uk/news/uk-wales-north-west-wales-25794295 (last accessed 26 April 2014).

54. 'Terrible railway accident in Wales', *The North-Eastern Daily Gazette*, 2 January 1883.

55. Jones, E. V. (1972), *Mishaps on the Cambrian Railways (1864–1922)*, Newtown, Montgomeryshire: Severn Press.

56. 'Railway accident in North Wales', *Western Mail*, 2 January 1883.
'Frightful accident', *The Dundee Courier and Argus*, 3 January 1883.

57. 'The railway accident in North Wales', *The Morning Post*, 3 January 1883.
'Extraordinary railway accident', *Lloyd's Weekly Newspaper*, 7 January 1883.

58. Gasquoine, C. P. (1922), *The Story of the Cambrian: A Biography of a Railway*, Oswestry: Woodall, Minshall, Thomas and Co. Ltd.

59. 'Railway accident in North Wales', *Western Mail*, 2 January 1883.
'Frightful accident', *The Dundee Courier and Argus*, 3 January 1883.
'The railway accident in North Wales', *The Morning Post*, 3 January 1883.
'Extraordinary railway accident', *Lloyd's Weekly Newspaper*, 7 January 1883.

60. Gasquoine, C. P. (1922), *The Story of the Cambrian*.
61. 'The railway accident in North Wales', *The Morning Post*, 3 January 1883.
62. 'The railway accident near Barmouth', *Leicester Chronicle and the Leicestershire Mercury*, 13 January 1883.
63. Gasquoine, C. P. (1922), *The Story of the Cambrian*.
64. 'Engine hurled over cliff', *The Times*, 6 March 1933.
65. 'Engine hurled over cliff', *The Times*, 6 March 1933.
66. 'Locomotive's fall into the sea', *The Times*, 10 March 1933.
    'Engine struck by landslip', *The Times*, 15 March 1933.
67. Jones, B. (1933), 'The geology of the Fairbourne – Llwyngwril district, Merioneth', *Quarterly Journal of the Geological Society of London*, 89, 145–171.
    Allen, P. M. and Jackson, A. A. (1985), *Geology of the Country around Harlech*, London: HMSO.
    Brenchley, P. J., Rushton, A. W. A., Howells, M. and Cave, R. (2006), 'Cambrian and Ordovician: The early Palaeozoic tectonostratigraphic evolution of the Welsh Basin, Midland and Monian Terranes of Eastern Avalonia', in Brenchley, P. J. and Rawson, P. F. (eds), *The Geology of England and Wales, 2nd Edition*, 25–74, Bath: Geological Society Publishing House.
68. Hawkins. T. R. W. (1985), 'Influence of geological structure on slope stability in the Maentwrog Formation, Harlech Dome, North Wales', *Proceedings of the Geologists' Association*, 96, 289–304.
69. Mount, A. H. L. (1933), Report for the information of the Minister of Transport, in accordance with the Order of the 6th March, the result of my Inquiry into the circumstances of the accident, which occurred at about 7.6 a.m. on the 4th March, in Vriog [sic] Cutting, between Llwyngwril and Fairbourne, near Barmouth, on the Cambrian section of the Great Western Railway, 11 pp.
70. BBC News (29 March 2005), 'Railway line closed by landslide', available at http://news.bbc.co.uk/1/hi/wales/north_west/4388583.stm (last accessed 26 April 2014).
71. Daily Post North Wales (31 March 2005), 'High tide blamed for rail landslide', available at http://www.dailypost.co.uk/news/north-wales-news/2005/03/31/high-tide-blamed-for-rail-landslide-55578-15351618/ (last accessed 26 April 2014).
72. Mourant, A. (2012), 'Time and tide', *Rail Professional,* March 2012, available at http://www.railpro.co.uk/magazine/?idArticles=1299 (last accessed 26 April 2014).
73. Ransom, P. J. G. (2001), *Snow, Flood and Tempest: Railways and Natural Disasters*.
74. Pearce, F. and Hamer, M. (1983), 'The Empire's last stand', *New Scientist*, 98, 364–367.
    Tilly, G. (2002), *Conservation of Bridges: A Guide to Good Practice*, London: Spon Press, 416 pp.
    Ransom, P. J. G. (2001), *Snow, Flood and Tempest: Railways and Natural Disasters*.
75. Tilly, G. (2002), *Conservation of Bridges: A Guide to Good Practice*, London: Spon Press, 416 pp.
    Ransom, P. J. G. (2001), *Snow, Flood and Tempest: Railways and Natural Disasters*.

76. Jehu, T. J. (1918), 'Rock-boring organisms as agents in coast erosion', *Scottish Geographical Magazine*, 34, 1–11.

77. Jehu, T. J. (1911), 'The glacial deposits of western Caernarvonshire', *Transactions of the Royal Society of Edinburgh*, 47, 17–56.

78. 'Result of landslip', *The Scotsman*, 21 February 1910.
'Severe gale', *The Times*, 21 February 1910.

79. 'Result of landslip', *The Scotsman*, 21 February 1910.

80. 'Havoc caused by the recent gale: A train disaster in Ireland', *Penny Illustrated Paper and Illustrated Times*, 26 February 1910.

81. Quale, H. (2007), *Whitehaven: The Railways and Waggonways of a Unique Cumberland Port*, Pinner: Cumbrian Railways Association, 101 pp.

82. 'The recent hurricane', *Liverpool Mercury*, 31 December 1852.
'Great wind, storms and tremendous flood', *The Preston Guardian*, 1 January 1853.

83. 'The recent storms', *The Morning Post*, 31 December 1852.

84. 'Extraordinary railway accident', *The Times*, 27 January 1852.

85. 'The recent storms', *The Morning Post*, 31 December 1852.

86. 'The recent storms', *The Morning Post*, 31 December 1852.

87. 'Great destruction at Whitehaven &c.' *Liverpool Mercury*, 2 February 1869.
'The severe storm and floods', *The Scotsman*, 2 February 1869.
'The gale', *The Times*, 3 February 1869.
'Great and destructive storm', *The Preston Guardian*, 6 February 1869.

88. 'Ferry aground as gale hits Britain', *The Times*, 28 February 1967.

89. BBC News Cumbria (6 January 2014), 'Repairs to storm-hit Cumbrian rail line "to take a week"', available at http://www.bbc.co.uk/news/uk-england-cumbria-25612478 (last accessed 26 April 2014).
BBC News Cumbria (13 January 2014), 'Storm-hit Carlisle to Barrow coastal rail line opens', available at http://www.bbc.co.uk/news/uk-england-cumbria-25710721 (last accessed 26 April 2014).

90. 'The recent storm, Serious damage to railways and agricultural stock', *The Scotsman*, 23 September 1891.

91. 'Opening of the South Wales Railway to Carmarthen', *The Morning Post*, 20 September 1852.

92. 'Opening of the South Wales Railway to Carmarthen', *The Morning Post*, 20 September 1852.

93. 'The late storm', *The Times*, 29 December 1852.

94. 'The violent gale', *The Times*, 24 January 1890.

95. 'Fatal gale and floods', *The Standard*, 9 October 1896.
'Gale & floods', *Liverpool Mercury*, 9 October 1896.
'Terrific gales', *The Dundee Courier and Argus*, 9 October 1896.

96. 'Fatal gale and floods', *The Standard*, 9 October 1896.

97. Millward, R. (2003), 'Railways and the evolution of Welsh holiday resorts', in Evans, A. K. B. and Gough, J. V. (eds), *The Impact of the Railway on Society in Britain*, 211–223, Aldershot: Ashgate Publishing.

98. Clark, E. (1850), *The Britannia and Conway Tubular Bridges with General Inquiries on Beams and on the Properties of Materials used in Construction*, Volume I, London: Day and Son, 466 pp.

99. Clark, E. (1850), *The Britannia and Conway Tubular Bridges*.

100. Ayres, G. (2011), *History of the Mail Routes to Ireland until 1850*, Lulu, 134 pp.
101. The National Library of Wales: Dictionary of Welsh Biography, Stanley Family of Penrhos, Anglesey, available at http://wbo.llgc.org.uk/en/s-STAN-PEN-1763. html?query=Stanley&field=name (last accessed 26 April 2014).
102. 'Accident on the Chester and Holyhead Railway', *The Huddersfield Chronicle and West Yorkshire Advertiser*, 29 October 1859.
    'Fearful catastrophe', *Freeman's Journal and Daily Commercial Advertiser*, 28 October 1859.
103. 'Loss of the Australian emigrant ship, "Royal Charter"', *Birmingham Daily Post*, 28 October 1859.
    Merseyside Maritime Museum, *The sinking of the Royal Charter*, available at http://www.liverpoolmuseums.org.uk/maritime/visit/floor-plan/emigration/ro yalcharter/ (last accessed 26 April 2014).
104. 'Disastrous gale', *The Scotsman*, 2 November 1887.
    'Holyhead Railway washed away', *Aberdeen Weekly Journal*, 2 November 1887.
    'Heavy gale at Chester', *Cheshire Observer*, 5 November 1887.
    'Terrific gale', *The Morning Post*, 2 November 1887.
    'The destructive gale', *The Pall Mall Gazette*, 2 November 1887.
    'Disastrous gale', *The Standard*, 2 November 1887.
105. 'Severe gale', *Daily News*, 2 November 1887.
106. 'The gale at Chester', *Cheshire Observer*, 10 October 1896.
107. 'More gale damage', *The Times*, 25 September 1945.
    'Engineers call for a new approach to sea defences', *The Times*, 1 March 1990.
108. 'Plea for government aid after £1m storm damage', *The Times*, 15 November 1977.
109. 'The violent gale', *The Times*, 14 January 1899.
110. 'Railway disaster on the Welsh coast: Train plunged into the sea', *The Scotsman*, 14 January 1899.
    'The disastrous gale', *The Standard*, 14 January 1899.
    'Railway disaster in North Wales', *Cheshire Observer*, 14 January 1899.
    Talbot, F. A. (1899), 'The constant rivalry of sea and shore', *The Windsor Magazine*, May 1899, 681–685.
111. 'Terrible disaster in Wales', *Daily News*, 14 January 1899.
112. 'Penmaenbach railway disaster', *The Morning Post*, 16 January 1899.
113. 'North Wales railway disaster', *The Times*, 24 January 1899.
114. 'North Wales railway disaster', *The Times*, 24 January 1899.
115. The Great Western Railway (1922), *Legend Land: Being a Collection of some of the Old Tales told in those Western parts of Britain served by the Great Western Railway, now retold by Lyonesse, Volume I*, London: The Great Western Railway, Paddington Station, 56 pp.
    The Great Western Railway (1922), *Legend Land: Being a Collection of some of the Old Tales told in those Western parts of Britain served by the Great Western Railway, now retold by Lyonesse, Volume II*, London: The Great Western Railway, Paddington Station, 56 pp.
116. Pennick, N. (1987), *Lost Lands and Sunken Cities*, London: Fortean Times, 176 pp.
117. Duck, R. W. (2011), *This Shrinking Land*.

118. Barham, R. H. D. (1867), 'The Monk of Haldon: A legend of South Devon', *Temple Bar*, 20, 488–496.
119. Robinson, A. and Millward, R. (1983), *The Shell Book of the British Coast*, Newton Abbot: David & Charles, 560 pp.
120. 'South Devon Railway', *The Morning Chronicle*, 28 October 1859.
121. 'Devonshire', *The Royal Cornwall Gazette, Falmouth Packet and Plymouth Journal*, 19 April 1844.
     Kay, P. (1990), *Rails along the Sea Wall*, Sheffield: Platform 5 Publishing Ltd, 60 pp.
122. 'The South Devon Railway', *The Times*, 31 October 1846.
123. 'The South Devon Railway', *The Times*, 21 November 1846.
124. 'Serious damage to the South Devon Railway', *The Standard*, 19 February 1855.
125. 'The storm and high tides', *Liverpool Mercury*, 2 February 1869.
     'The severe storm and floods', *The Scotsman*, 2 February 1869.
     'The storms and floods', *The Morning Post*, 3 February 1869.
     'The gale', *The Times*, 3 February 1869.
126. 'Fresh inroads of the sea on The South Devon Railway', *The Huddersfield Daily Chronicle*, 1 January 1873.
127. 'Subsidence on G. W. R.: Breach in sea wall near Dawlish', *The Times*, 6 January 1930.
128. 'More mines washed ashore', *The Scotsman*, 24 December 1945.
129. Kay, P. (1990), *Rails along the Sea Wall*.
130. Ransom, P. J. G. (2001), *Snow, Flood and Tempest: Railways and Natural Disasters*.
131. Kay, P. (1990), *Rails along the Sea Wall*.
132. Durrance, E. M. (1969), 'The structure of Dawlish Warren', *Proceedings of the Ussher Society*, 2, 91–101.
133. Collin, R. L. (1998), 'A digital map on the Internet: Dawlish Warren, Devon', *Earth Surface Processes and Landforms*, 23, 1269–1271.
134. Durrance, E. M. (1969), 'The structure of Dawlish Warren'.
135. Pye, K., Saye, S. and Blott, S. (2007), *Sand Dune Processes and Management for Flood and Coastal Defence. Part 4: Techniques for Sand Dune Management*, Joint Defra/EA Flood and Coastal Erosion Risk Management R&D Programme, R&D Technical Report FD1392/TR, 49 pp.

3

The Edge of Collapse

Earl of Gloucester
Dost thou know Dover?

Edgar
Ay, master.

Earl of Gloucester
There is a cliff, whose high and bending head
Looks fearfully in the confined deep.
Bring me but to the very brim of it,
And I'll repair the misery thou dost bear
With something rich about me. From that place
I shall no leading need.

King Lear, Act 4, Scene 1

Coastal landslides

For a relatively small island – or cluster of islands – Britain has a huge variety of rock types. This geological diversity gives rise to a wide range of landforms and scenery around the coastal edge. Between the extremes of high, precipitous cliffs, carved out of durable rock formations, to low-lying sandy dune fields or mudflats, the vast array of coastal types is striking. Landslides – movements of masses of rock, earth or debris down a slope[1] – are perfectly natural phenomena along our coasts where cliffs are present, but the magnitude and frequency of such events vary according to the local geological conditions. At the coast, landslides are triggered mainly by wave erosion undercutting the base of a cliff. However, the infiltration of rain water from above can also play an important role in cliff destabilisation, as can ground motion caused, for instance, by earthquakes or even the passage of heavy trains. The stability of rocky cliffs is not only a function of the rock type or types present but importantly the orientation of and the spacing between the various planes of weakness – bedding planes, joints and faults; collectively known as discontinuities – that sub-divide the rock mass. In materials that lack well defined systems of discontinuities, the mode of failure is typically on a concave upwards, curving surface; this results in what is known as a rotational landslide that may develop into a slump or debris flow onto the beach at the foot. The final trigger may perhaps be a heavy fall of rain that saturates and increases the weight of the material above the slip face or undercutting by wave attack at the toe or indeed a combination of these. Our coastal railway lines have thus been constructed, often precariously, on top of, at the foot of or, by means of tunnels, through a multiplicity of earth materials on the edge. Moreover, these materials vary in their stability. It is also noteworthy that landslide-prone, unstable terrain is often of very high scenic value and of considerable aesthetic appeal.

Sweet Poppyland

On 17 July 1898, the popular Sunday broadsheet *Reynolds's Newspaper* announced to its readers that Mr Clement Scott had been bitten by a dog and was therefore unable to travel: 'The doctors tell me to rest and take things easy until the dog is proved to be mad or sane.'[2] Directly beneath this notification the paper reported that another large landslide of the Lighthouse Hills Cliffs had occurred to the east of Cromer, near to the same location as one that had taken place about two months previously: 'This time a piece of the cliffs 30 yards wide by 120 yards long has been hurled onto the beach, where no small portion of the first slip yet remains, despite the action of the sea.'[3] The newspaper did not apparently appreciate that the link between these two stories was much more than their mere juxtaposition at the bottom of a column headed *Evils We Might Avoid.*[4]

Fifteen years earlier, Clement Scott had created 'Poppyland'; long before the red bloom had become Britain's flower of remembrance. Scott was a London journalist, author, poet and theatre critic[5] who first introduced readers of *The Daily Telegraph* to this place of 'solitude, fine air, scenery, and seclusion' on 30 August 1883.[6] Making reference to the teeming abundance of the red flowers in the area, he used the term to refer to the cliff-top lands to the east of Cromer in Norfolk. Susceptible to land sliding and vulnerable to erosion by the sea, the cliffs at Cromer, Overstrand and Sidestrand are highly unstable, sometimes spectacularly so. Indeed Cromer has endured a long battle with the sea for centuries, which it has frequently lost.[7] The then small settlement was severely damaged in a storm in 1836. This was the catalyst for the construction of a sea wall and eventually a promenade, which have been extended and reinforced on many occasions as Cromer expanded and gained prominence as a watering place.[8] The cliffs that characterise this stretch of coast are composed of unconsolidated till laid down as the ice retreated and melted at the end of the Pleistocene glaciations of the region. Where these materials are unsupported by a retaining wall, as beyond the eastern and western limits of the town, successive rotational landslides on curving surfaces of failure have created small valleys, depressions and scars in the cliff top above. It was from within this terrain that Scott looked westwards to Cromer (Figure 3.1):

I turned my back on perhaps the prettiest watering place of the East Coast and walked along the cliffs. At a mile removed from the seaside town I had left I did not find a human being. There they all were below me, as I rested amongst the fern on Lighthouse Cliff . . .[9]

Figure 3.1 The view along the coast westwards towards Cromer over terrain, known as Happy Valley, much disturbed by landslide activity. Much of the 'Poppyland' that Clement Scott knew has now slipped into the sea (photo: R. W. Duck).

The Great Eastern Railway had reached Cromer in 1877 and it was by that means that Scott had travelled from London six years later. He was to become immediately enamoured by that part of north Norfolk's coastal edge. The Great Eastern Railway Company almost instantly embraced the theme of Poppyland, helping to elevate it to legendary status, and newspapers of the day did their bit to entice and encourage the would-be traveller:

> The strenuous efforts of the Great Eastern Railway Company these late years to attract custom have not been without their effect upon the popular mind. There are cheap excursion bookings on Friday to all the places of repute, in tourists' estimation, in the Eastern counties; you may also travel to Scotland under their aegis. On Saturday one may travel at a reduced rate to sweet Poppyland – to Cromer and Mundesley – or explore the wondrous delights of the Broad district, or penetrate perchance, the unromantic-looking Fens.[10]

Indeed, the creation of Poppyland was to have a huge impact on the area, drawing Victorian tourists to the watering place in their thousands by rail, something Scott was to ultimately lament: 'The Cromer that we visit now is not the Cromer I wrote about but a few short years ago as my beloved Poppyland.'[11] In fact Scott deplored vehemently the developments of 'bungalow land', for which he shouldered the blame, that had taken place since he popularised Poppyland. He berated himself: 'What a fool you were ever to let a human being into the secret of your beloved Cromer. Why on earth did you not keep it to yourself?'[12] Some local people certainly agreed with Scott. On 25 August 1891, a correspondent to *The Pall Mall Gazette*, one Mr F. Henderson of Overstrand, protested passionately that Scott's popularisation of Cromer had led to widespread and unwelcome housing developments on an uncontrolled scale: 'The speculative builder has made it his own; but we did hope that Poppyland would be left to us.' Poppyland was once a sacred spot but, 'thanks to the London newspapers dire sacrilege has been committed; and Poppyland has gone the evil way of Cromer.'[13] In a barbed reference to Clement Scott, Mr Henderson continued, 'No poet will sing the praise of Overstrand again. This once lovely village is overgrown with prim residences – *desirable mansions*, as the advertisements put it.'[14] Moreover, Scott's romantic appeal along with the ease of access afforded by the railway had led to a serious decline in the 'type' of person that visited the area:

I have today seen in the peaceful old Garden of Sleep a party of Cockneys sitting around the remains of a luncheon and singing a ribald song. And yet a man is supposed to keep his temper and refrain from swearing in a decent society.[15]

Completely incandescent, Mr F. Henderson of Overstrand concluded acerbically, 'I am afraid any excuse is that I declined to regard such society as decent.'[16] Less than a week later, on 31 August 1891, the newspaper published a riposte from one Alfred Berlyn of the same village who asserted that Mr F. Henderson of Overstrand regarded Poppyland through 'jaundiced magnifying-glasses.'[17] In those days, long before political correctness was even heard of, Mr Berlyn concluded that:

We have no pier [the present structure was to be built in 1901], no parade, no niggers, no bathing-machines and no *cheap-trippers*; and I unhesitatingly assert that Poppyland remains as charming, as restful and as beautiful to-day as when I first knew it several years ago.[19]

Regardless of these opposites on the spectrum of opinion, natural processes were unremitting and the soft cliffs of the north Norfolk coast continued to succumb to attack by the sea.

One of Scott's fiercest critics, known only as 'H.', remarked of Cromer in *The Saturday Review* that, 'Only the sea remains inviolate; upon everything else the hand of the spoiler has been cruelly heavy.' In an increasingly bitter personal attack, 'H.' continued:

In an evil moment some dozen years ago that egregious person Mr Clement Scott discovered the district. That was the beginning of the end . . . He labelled it 'Poppyland' – the sort of thing a cockney would do and think pretty poetical.[19]

'Sweet Poppyland' was, in fact, located on the top of one of the most unstable and rapidly eroding cliffs in England – and this was a key attraction to the excursionist. The railway companies in the 1920s and for two decades later even made a virtue in imagery – albeit perhaps unwittingly – of the natural instability of this area. A classic poster of the 'Gem of the Norfolk Coast' depicted the unstable cliff edge to the east of the town along with people sitting upon the scarred surface of the cliff top, created by the slipping of the unsupported land towards the sea. Just a few red poppies are visible, some growing on displaced masses of turf (Figure 3.2).

Figure 3.2 'Cromer – Gem of the Norfolk Coast': poster produced for the London Midland & Scottish Railway and the London & North Eastern Railway, 1923–47; artwork by Walter Dexter. The poster depicts the unstable cliffs of the Cromer area and shows evidence of recent land sliding. People are shown either sitting on seats or standing close to the edges of the slipped masses of earth and, in particular, two women in the foreground are shown sitting on the grass on top of a slipped mass with their legs dangling over the edge (Image No. 10170751, courtesy of the National Railway Museum/Science & Society Picture Library).

A striking man-made feature within the solitude of Poppyland was a round tower perched on the edge of the cliff: 'Not desiring to be followed I strolled on, attracted by a ruined church tower, took a cut through the cornfields towards a cluster of farms and a distant village.'[20] This had once been part of St Michael's Church, Sidestrand, but, because of its precarious location and the risk of imminent collapse due to land sliding and undercutting by the sea, it had been demolished leaving the tower in isolation. Tower Lane, leading to the cliff edge is today a reminder of the structure's existence. In fact, this was not the original tower; its predecessor is said to have collapsed on a stormy night in 1841. The tower of Scott's day had been built in 1848.[21] Amidst the poppies he called it, and the surrounding graveyard of the former parishioners of Sidestrand, 'The Garden of Sleep' and his romantic poem *A Summer Song*, along with photographs of the tower, became the subjects of postcards. 'H.' was less than impressed with what he called 'a maudlin sentimental song' that 'was squalled for years in suburban drawing-rooms'.[22] The tower and garden were to feature in many of Scott's other works. Even after its inevitable fall into the sea, postcards showing the tower continued to be published until the 1930s to intrigue the visitor – carrying the footnote, 'The Tower Fell in 1916' (Figure 3.3). Scott, however, did not survive to see this spectacle – he had passed away in 1904. A photograph of the seaward-leaning tower[23] – on the brink of collapse into the sea – clearly reveals, like so many of the traditional buildings in Cromer and the surrounding area, that it was faced with well-rounded cobbles and pebbles, principally of chalk and flint. These had been locally sourced by its builder from the beach below. The graveyard was also to succumb to the sea and human bones sticking out of the cliff or lying on the beach below, along with remnants of the tower, were a spectacle of macabre fascination to the Victorian visitor. Much of the actual land trod by Scott has now fallen into the sea. Here, in common with elsewhere in Britain, today's modern farming practices and associated habitat destruction have greatly reduced the number of poppies in the fields[24] and the mown cliff-top fairways of the Royal Cromer Golf Course now dominate the area, taking strategic advantage of topographic undulations created by land rotating and sliding towards the sea. Needless to relate, the golf course itself has had to be realigned periodically over the years as wasting of the cliffs has ripped out tees, parts of fairways and greens.[25] A water trough memorial beside the road from Cromer to Overstrand, today located very much nearer to the sea than at the time of its erection, bears the inscription:

To CLEMENT SCOTT,
WHO BY HIS PEN IMMORTALISED 'POPPYLAND'
ERECTED BY MANY FRIENDS NOVEMBER 1909.

By contrast and perhaps not surprisingly, 'H.'s final words on Scott were less gracious: 'Was there not Brighton, or even Margate, that Mr Scott must needs disport himself in Norfolk?'[26]

The link between the railway and Poppyland – and without the railway Scott could never have made it so popular – lives on, albeit in a slightly different location from that of his original creation. The North

Garden of Sleep, near Cromer.

POPPY-LAND.
A PRELUDE.—THE GARDEN OF SLEEP!

ON the grass of the cliff, at the edge of the steep,
    God planted a garden—a garden of sleep!
'Neath the blue of the sky, in the green of the corn,
It is there that the regal red poppies are born!
Brief days of desire, and long dreams of delight,
They are mine when my Poppy-Land cometh in sight.
In music of distance, with eyes that are wet,
It is there I remember, and there I forget!
O! heart of my heart! where the poppies are born,
I am waiting for thee, in the hush of the corn.
        Sleep! Sleep! From the Cliff to the Deep!
        Sleep, my Poppy-Land, Sleep!

In my garden of sleep, where red poppies are spread,
I wait for the living, alone with the dead!
For a tower in ruins stands guard o'er the deep,
At whose feet are green graves of dear women asleep!
Did they love as I love, when they lived by the sea;
Did they wait as I wait, for the days that may be?
Was it hope or fulfilling that entered each breast,
Ere death gave release, and the poppies gave rest?
O! life of my life! on the cliffs by the sea,
By the graves in the grass, I am waiting for thee!
        Sleep! Sleep! In the Dews by the Deep!
        Sleep, my Poppy-Land, Sleep!

Clement Scott.   Copyright.   Jarrolds.
Tower Fell, 1916.

Figure 3.3 Postcard from the 1920s depicting the tower in the 'Garden of Sleep, near Cromer' accompanied by Clement Scott's romantic poem or his 'maudlin sentimental song', depending on one's point of view.

Norfolk Railway, a section of preserved heritage steam railway to the west of Cromer between Sheringham and Holt, brands itself as the 'Poppy Line'. At Poppyland the link between coastal landslide terrain and railway was romantically platonic. Elsewhere at the edge of Britain the relationship was far more physically intimate.

'The Town that Never Was'

The coastal railway between Scarborough and Whitby via Robin Hood's Bay in North Yorkshire opened in 1885. This inter-watering place route, another Dr Beeching casualty in 1965, traversed some truly spectacular country:

> As is well known the Scarborough and Whitby line runs along the most rugged piece of coast country in the north of England – over hills and dales, across ravines and water-ways, through tunnels and mountain streams. It skirts beetling cliffs and roaring cataracts, and bends round twisted curves a boa-constrictor would find difficulty in negotiating. Below is the sea on a rock-bound coast some hundreds of feet, and sheer above rise hills uncompromising in their ruggedness . . . No country was ever more unlikely for successful railway engineering, and no line in England cost more to construct.[27]

It was remarkable for its very steep gradients of up to 1 in 39 at Ravenscar and nearby.[28] Whitby, Robin Hood's Bay and Scarborough are no strangers to landslides and, similarly, the cliffs in between (Figure 3.4). Composed largely of a variety of sedimentary rocks of Jurassic age, principally sandstone and very friable shale, these are prone to instability, especially after heavy rainfall.[29] The line was thus perched along a rather unstable edge, which, for much of its route, contributed to its beauty. An early guide, published shortly after it had opened, noted that:

> The line has opened out the country and made it accessible in all its virgin loveliness. The line runs through pleasant undulating pasture lands at either end, winds in and out among the gorse and heather clad hills, dips into wooded dales, skirts the edge of a wild moor, climbs the highest cliff on the Yorkshire coast, runs around one of the bonniest bays in the Kingdom, and over a portion of its course is perched on the brow of a cliff against which the waves ceaselessly break.[30]

Figure 3.4 Looking north from Ravenscar along the landslide-prone terrain through which the Ravenscar to Robin Hood's Bay section of the Scarborough to Whitby railway once passed. Today the old track bed, seen passing beneath the over bridge, is a footpath and cycleway (photo: R. W. Duck).

And, needless to say, it was breached from time to time as the unstable rocks beneath it gave way and slipped seawards. For instance, a series of serious landslides, principally of shale, took place roughly mid-way between Ravenscar and Robin Hood's Bay over a period of several days in early August 1888. The line, opened just three years earlier, was obstructed for many days. Shortly after the first landslide a huge crack appeared near the brow of the hill above. This gradually opened up until many thousands of tons of rocky debris toppled over; completely burying the line.[31] Similarly, in late February 1900, following a period of prolonged heavy rainfall, another series of landslides took place. A northbound train from Scarborough passed through without incident but the passage of the next southbound service was blocked completely by a large quantity of fallen earth at a site that had experienced such events on many occasions previously.[32] Several thousands of tons of debris had, yet again, slipped onto the track, over a length of about a quarter of a mile. But this was not all. Perched on the cliff edge, the rain waters had undermined the sleepers and cut a chasm 25 feet deep below the track with the result that a stretch of about 40 yards in length had subsided.[33] Snow falls during the few days leading up to Christmas 1925 in the area were followed by melt water floods. Culverts beneath the line were unable to cope with the volume of runoff and, as a result, about 30 metres of the single track formation were washed out.[34]

For such a small village, Ravenscar, the highest point on the line, had an unusually long station platform, a feature that still remains long after the last train departed. A grand proposal was made in the mid-1890s to turn this sleepy cliff-top village into a resort, a watering place to rival nearby Scarborough.[35] It is difficult to envisage how this might have happened owing to the lack of a sandy beach and the difficulty with which the rocky shore below can be accessed. The cliff-top platform was, however, built in anticipation of lengthy excursion trains and is now a hollow reminder of 'The Town that Never Was'.[36] Although many plots of land were sold, roads and drains laid, it is providential that this development never came to fruition given the instability of the cliffs along this particular section of the coast.

Folkestone to Dover: landslides and more

Few if any railway travellers on today's high speed Javelin Train between the English Channel ports of Folkestone and Dover will realise that they are passing through what is naturally some of the most unstable coastal terrain in the country. Opened in 1844 by the South-Eastern Railway Company,[37] the line negotiates its way through a 3-kilometre-long series of very large and deep-seated landslides – an area of undercliff known as Folkestone Warren. The railway lines thread through a series of three tunnels very near the sea along East Wear Bay: closest to Folkestone, the Martello Tunnel (580 metres in length), Abbot's Cliff Tunnel (1.7 kilometres) and, closest to Dover, Shakespeare Cliff Tunnel (1.3 kilometres). The Channel Tunnel from Folkestone to Calais passes obliquely beneath the line of the latter. To the west, along the edge of the open railway between Abbot's Cliff and Shakespeare Cliff Tunnels is an area of claimed land, a country park known as Samphire Hoe. This platform was built from five million cubic metres of chalk marl spoil, the debris excavated from the Channel Tunnel; landscaped to cover an area of 36 hectares and bounded from the sea by a vertical wall almost 2 kilometres in length.[38] Whilst Samphire Hoe is principally a public amenity, it also provides an important protection to this section of the Folkestone to Dover railway from coastal erosion.

On leaving Shakespeare Cliff Tunnel to the east, the line, as originally built, arrived 'at a loose shingly beach on which the sea continually beats, and in rough weather with great violence.'[39] It was thus carried towards Dover on a wooden viaduct: 'A sea wall would not in this instance have served the purpose; the sea would have washed it away.' It was further suggested that a solid structure would have failed along the shore at this location, whereas 'a light open timber-work framing, carrying the rails on an elevated platform, has been found to answer every requirement.' It is perhaps of little surprise to find that today a 'solid' sea wall carries the railway. The assertion at the time of the opening of the line that, 'A wooden viaduct exposed to the fury of a south-wester [sic], is in fact as safe as Waterloo Bridge',[40] was soon proved to be rather too bold (Figures 3.5 and 3.6).

The bulk of the precipitous cliffs – the White Cliffs of Dover – are composed of chalk, of Cretaceous age and relatively permeable to

Figure 3.5 Looking towards Dover from the eastern end of Shakespeare Cliff. Here the former wooden viaduct carrying the railway from Folkestone has been replaced by a sea wall built on land claimed from the beach (photo: R. W. Duck).

Figure 3.6 The twin eastern portals of Shakespeare Cliff Tunnel, Dover. Recent falls of chalk from the cliff are evident (photo: R. W. Duck).

rainwater. Beneath these rocks, at the base of the cliff, lies the impermeable Gault Clay Formation, which, in turn, rests upon permeable sandstones of the Folkestone Formation. As in other locations with similar geology in the south of England, such as Ventnor on the south coast of the Isle of Wight, land sliding takes place at the boundary between the chalk and the clay or between clay and sandstone. The clay layers soak up rainwater percolating downwards and can act as slip surfaces. When lubricated by water, seaward mass movements can take place along bedding planes, often accompanied by back-tilting rotation of the slipped masses. It is through this displaced and disturbed terrain that the railway passes. Landslides, however, took place long before the railway was built, with records going back to the eighteenth century. The first documented landslide took place in 1765.[41]

During the course of the construction of the line and its tunnels, storms in October 1841 greatly reduced the level of the beach at Dover, undermining boathouses and other buildings and stripping away large amounts of sand and shingle. In the previous year an enormous rock fall of chalk had taken place to the west of Dover at Round Down Cliff.

Some people believed that the fallen debris had arrested the longshore transport of shingle towards Dover thereby contributing to the lowering of the beach level, the natural supply of sediment having been depleted. Another view was put forward that this was the result of the direct removal of beach materials to the east of the site of Round Down Cliff for the construction of a sea wall to defend the line between Dover and Folkestone. It was noted in *The Times* that, 'the presence of the beach was a great preservative to the town, while its absence has been the sole cause of the late destruction to property.'[42] In fact, virtually the whole of Dover beach was removed such that, 'nothing now remains to prevent the sea washing against that vast and stupendous cliff which Shakespeare has immortalized.'[43] The implication that the South-Eastern Railway Company might be responsible for the loss of Dover beach prompted a speedy response from its Secretary who vehemently denied the allegation that this was due to shingle extraction for the construction of its sea wall.[44] Less than two years later, on 26 January 1843, Round Down Cliff – between the eastern portal of Abbot's Cliff Tunnel and the western portal of Shakespeare Cliff Tunnel – was removed by blasting, on this occasion to facilitate the laying of a straighter line of railway:

> You will not be surprised to hear that the annunciation that an explosion of 18,000lb. of powder was to be made in the Round Down Cliff this afternoon brought an influx of strangers into this town; still, though considerable, it was not as large as I had expected. Curiosity was, I think, paralysed by a vague fear of danger, which kept some thousands at home who might have witnessed it, as the event turned out, without the slightest shock to their nervous system.[45]

The success of the operation – over in 4 or 5 minutes and hailed as a splendid engineering triumph[46] – ensured that the line would be open earlier than some had forecasted. It was reported that the total cost of the operation was £500 and for the South-Eastern Railway Company to have used any other means to remove the cliff would have cost fully £6,000:[47] A mass of rock weighing one million tons was said to have been removed by the blast.[48]

> Exactly at twenty-six minutes past two o'clock a low, faint, indistinct, indescribable, subterranean rumble was heard, and immediately afterwards the bottom of the cliff began to belly out, and then

almost simultaneously about 500 feet in breadth of the summit began gradually, but rapidly to sink . . . The rock seemed as if it had exchanged its solid for a fluid nature, for it glided like a stream into the sea . . . filling up several large pools of water which had been left by the receding tide.[49]

Local people were fully aware of the history of land sliding along this coast and many were sceptical of the official report into the railway's safety and stability prior to its opening. Major-General Pasley, Inspector-General of Railways, 'having inspected the whole of the ground above the railway with the greatest attention, in order to discover the unsound parts of the chalk cliff, if any, which may be known by cracks at the surface,' was completely satisfied that there was no danger to the travelling public. 'I am of the opinion', he wrote on 1 February 1844, 'that there is not the smallest ground for apprehension', either with respect to the stability of the tunnels along the route or the portions of the line 'formed along the beach.'[50] In the case of the latter, it was noted that the installation of strong sea walls prevent them from being damaged by storm waves. Major-General Pasley concluded his findings with unequivocal prose:

Upon the whole I have great pleasure in assuring your Lordships not only that the railway itself is in a perfectly safe and efficient state, but that no part of the works are exposed to the smallest danger, either from the interruptions of the sea or from the fall of the cliffs; though it was natural for the public to have their doubts, in the first instance, as to the success of so very arduous an undertaking.[51]

There was, in the Major-General's opinion, no need whatsoever for local people or indeed others to be concerned:

As everything necessary for the public safety has already been done, for I have repeatedly passed over both lines of rails at full speed on a special engine or in the carriage attached to it, I beg leave to recommend that your Lordships will be pleased to authorize the directors of the South-Eastern Railway Company to open their line to Dover as soon as they shall have erected a temporary station for the accommodation of passengers, as they propose, at that place, and which will probably be finished in a few days.[52]

The requisite station was duly assembled with the predicted alacrity and, amidst great ceremony, the line opened within a week of the publication of the Inspector-General's report.[53] Major-General Pasley, however, was to be proved wrong. Indeed many natural incidents were to subsequently interrupt rail traffic on the line, most spectacularly so in 1915.

In January 1877, a series of landslides and rock falls in Folkestone Warren following a period of heavy rainfall interrupted railway traffic.[54] The railway was blocked for about 100 metres and one of the railway company's workers was killed instantly in one of the falls of chalk.[55] The rails were buckled and there was serious damage to Abbott's Cliff Tunnel. It was two months before a train, carrying Sir Edward Watkin, Chairman, and several Board members of the South-Eastern Railway Company, could once more pass along the line.[56] Although this was a serious event, it was minor compared with the landslides of 1915 in Folkestone Warren. The year had been extraordinarily wet and especially so in the final four months. On 19 December an exceptionally large landslide began to develop and this ultimately involved the greater part of the Warren in a bodily movement towards the sea. The maximum displacement was about 50 metres near to the centre of the disturbance. Shortly after 6.00 p.m. a watchman noticed that there had been a subsidence in the line at the Martello Tunnel end of the Warren. He gave his red light to a group of nearby soldiers to stop a train expected at 6.10 p.m. from Ashford to Dover and went to raise the alarm by telephone. Soon after, the landslide proper began to develop on a rotational surface – the railway line became buried under about 12 metres of debris along a length of over 200 metres.[57] Shortly afterwards – at about 6.45 p.m. the train to Dover left Folkestone running late. There were about 130 passengers on board. On emerging from the Martello Tunnel and seeing the soldiers and the red light, the driver applied the brakes but the train became caught up in the subsiding ground as the slide continued to rotate. The engine and front carriages sank slowly as the rails became buckled and in this disabled state the train came to a halt. The passengers were obliged to make their way back to Folkestone on foot, through the Martello Tunnel.[58] Such was the extent of the damage to the railway that it was closed to traffic for almost four years, not opening again until early August 1919.[59]

Since the events of 1915, an extensive programme of monitoring, land stabilisation and drainage has been carried out, along with coastal protection. Deep test boreholes were drilled in the area in 1938 following another set of major landslides in 1936–7, which moved the western third of Folkestone Warren towards the sea by about 30 metres. These boreholes confirmed that the landslides had arisen by shear failure in the base of the Gault Clay, close to the top of the sandstone beneath. Erosion by the sea – removing material at the toe of the landslides – had exacerbated the rotational movement. Today, sophisticated networks of drains have been installed to lower the water table within the unstable cliffs of the Warren and concrete sea walls along with rock armour aprons provide protection from wave erosion at the foot.[60]

It is possible that the South-Eastern Railway Company was itself, at least in part, responsible for the landslide-induced damage to this section of their line through Folkestone Warren. The Company had already been implicated in beach erosion at Dover during the building of the line but it was their activities at Folkestone that have been brought into the spotlight. The predominant direction of longshore sediment transport along this coast is towards the east, which is towards Dover from Folkestone. Following the acquisition of the Harbour Pier at Folkestone, the South-Eastern Railway Company, as noted in the previous chapter, embarked on a series of progressive extensions to the structure. These took place in three phases; 1861–3, 1881–3 and finally 1897–1905.[61] As each extension was completed, it led to increasing obstruction of the natural sediment transport along the coast, causing progressively more material to be prevented from passing the pier. Thus, sand and shingle accumulated on the western side, whilst downdrift beaches to the east became ever more depleted in their supply. This effect was noted in 1883 after the second of the three phases of extension:

> Beautiful as the Warren still is, it is not to be compared with the Warren of half a century ago. The ingenuity of our present generation, which has constructed a harbour where our forefathers utterly failed, has not been without effects which lovers of the grand and beautiful scenes of nature must regret. While the beach has accumulated to an enormous extent west of the harbour, the rugged form of Copt Point has crumbled away until nothing is left but the rocky foundation on which it stood, and year by year the Warren which nestled behind it grows less and less in extent, as the restless waves roll up and swallow it.[62]

The reduction in supply of sediment removed the natural protection from wave attack and thus erosion became far more prevalent at the foot of Folkestone Warren. During the nineteenth and early twentieth centuries, from the initial construction of the pier in 1810, until 1915, there were eight significant phases of landslide activity. It has been suggested that the successive extensions of Folkestone Harbour Pier, particularly between 1810 and 1905 have induced a progressive increase in landslide activity in Folkestone Warren, which culminated in the 'great slide' of 1915.[63] Although there has been subsequent landslide activity – as in 1936–7 – the improved sea defences and drainage measures have greatly enhanced the stability of the cliffs. It is thus likely that the South-Eastern Railway Company might well have been responsible for indirectly triggering the landslides that led to the extensive and loss-making closures of its own line. The South-Eastern may have also initiated another, even more unusual and unpredictable phenomenon at Folkestone.

At 8.18 a.m. on 28 April 2007 Folkestone and Kent were rocked by an earthquake. In parts of Folkestone, close to the epicentre, many houses suffered extensive damage to chimneys and walls. The event, of local magnitude ML = 4.2, was by no means unusual for Britain; however the intensity of shaking and the level of damage to buildings have rarely been exceeded since records began in the country.[64] Although the jury is very much out, a somewhat controversial suggestion has been made that this was not a natural occurrence, but was induced by human activity – land gain by the build-up of shingle at Folkestone – since the construction of the Harbour Pier began in the early nineteenth century. The increased loading on the Earth's crust could, it has been suggested, have caused the April 2007 earthquake, by which time an estimated 2.8 million tonnes of sand and gravel had accumulated to the west of the pier.[65] If the seismic theory was indeed true, the South-Eastern Railway Company might not only have been responsible for causing increased coastal erosion and for triggering railway-disabling landslides, but also for ultimately bringing about earthquake-induced damage to the town of Folkestone. Furthermore, earthquakes aside, the related activities of the South-Eastern Railway Company, principally under the Chairmanship of the entrepreneurial Sir Edward Watkin, were to be the direct cause of other problems on the Kent coast, as will be explored in Chapter 5.

Under the edge

Between Dawlish and Teignmouth, as described in the previous chapter, the South Devon Railway, built on top of the sea wall, occupied a narrow confine, which was once part of the beach between the cliffs and sea. The friable nature of the Permian sandstones is such that rock falls and landslides from these cliffs have always been a major problem and a frequent occurrence. The days leading up to Christmas of 1852 were marked by terrible weather along the western and southern coasts of Britain. The railway embankment at Llanelli had been breached just two months after the line had opened, the Cumbrian coast line had been severely damaged in several places by storm waves (see Chapter 2) and, on Boxing Day 1852, a large landslide disrupted the South Devon Railway. Fearing for the safety of the line following the recent heavy rain and storm-force winds, watchmen had been stationed along the route to give warning of imminent danger. Prompt action halted the passage of a morning train from Dawlish before it would have ploughed into the 'most tremendous fall of cliff' that took place close to the Dawlish end of Parson's Tunnel near Holcombe, thus averting a possible derailment.[66] No sooner had the debris been cleared from the rails and re-opened to traffic, when another landslide occurred at the same spot, two days after the first. Again the line was completely blocked between Teignmouth and Dawlish and the occurrence of this second landslide impelled Brunel, along with several directors of the railway company, to make a site visit and superintend the removal of the fallen mass.[67] Fortunately, no life was lost. Just over three months later, on 4 April 1853, a potential tragedy was again averted thanks to the quick action of an engine driver. A locomotive pulling a passenger train ran off the line and was almost precipitated into the sea. Had not the driver turned off the steam at the very moment which he did, it is likely that all of the passengers on board would have been killed. The accident again occurred along the section of the wall between Clerk's Tunnel and Parson's Tunnel – named after the nearby, prominent sea stacks of the Parson and Clerk Rocks – close to the site of the December 1852 landslides. It transpired that the engine had been wrongly diverted along temporary rails, laid as a siding to enable material to be removed from the unstable cliffs. These rails should have been removed prior to the passage of the train but were not. The consequence was that the engine ran off the main line and into the debris adjoining it and when it stopped it was on the brink of falling over the edge of the sea wall.[68] Landslides and rock falls in this area became such a frequent occurrence that, by the early 1920s, Parson's Tunnel had been artificially extended northwards by means of brick and concrete[69] to prevent debris from falling onto the rails and blocking the passage of trains.

Nearly 150 years before the Folkestone earthquake an event of similar magnitude took place on 6 October 1863, with its epicentre close to Hereford in Worcestershire. This locality has been subject to several such occurrences over the years and the effects of this particular quake were felt at numerous places over the southern and central parts of England.[70] It elicited – as was common at the time – a whole host of letters to *The Times* over the next few days, including one on 8 October from Charles Dickens, then resident near to Rochester in Kent:

> I was awakened by a violent swaying of my bedstead from side to side, accompanied by a singular heaving motion. It was exactly as if some great beast had been crouching asleep under the bedstead and were now shaking itself and trying to rise. The time by my watch was 20 minutes past 3, and I suppose the shock to have lasted nearly a minute. The bedstead, a large iron one standing nearly north and south, appeared to me to be the only piece of furniture in the room that was heavily shaken. Neither the doors nor the windows rattled, though they rattle enough in windy weather, this house standing alone, on high ground, in the neighbourhood of two great rivers.[71]

The previous day, however, *The Times* had published a short but intriguing letter penned by 'Your obedient servant, J. T. T.' of Exeter:

> Sir, – Between 3 and 4 o'clock this morning the shock as of an earthquake was felt in this city and neighbourhood. Some persons were much alarmed and got up and examined their premises. In my own house, which is situated in a central part of the city, none of the inmates beside myself and wife noticed it. The bed shook under us. The vibration was similar to that caused by a railway train passing near, and such a sensation I have often experienced at Dawlish where the railway is near the dwellings.[72]

With its combination of the passage of heavy trains beside or beneath soft rocks right at the coastal edge, was there a fatal accident waiting to happen at Dawlish? Though rock falls were common and, without doubt, exacerbated by vibrations caused by rail traffic, it was to be over two decades after its opening that the greatest tragedy on this spectacular stretch of coastal railway took place. An initial account of the events of 29 August 1885 reported:

> A dreadful accident occurred at Dawlish this morning, a large portion of rock falling from the cliffs and burying a number of

people. Two ladies and two lads have been taken out alive, having escaped death in a miraculous manner. The bodies of a lady and a baby have also been taken out dreadfully crushed. It is believed that others are still lying under the falling rocks.[73]

As the details of the accident became clearer, it emerged that on this August Saturday at mid-day seven members of the family and household of Lady Graves-Sawle of Honiton had made their way to a picnic spot at the foot of an overhanging cliff on the Dawlish side of the southern portal of Kennaway Tunnel (Figure 3.7). This was a popular place, which was frequented every morning by many hundreds of local bathers as well as those who had come to Dawlish by train from Exeter. The friable rocks of the cliffs were widely known to be dangerous in this area and a notice warning strangers of rock falls had, ironically, 'been carried away by a landslip not two months before.'[74] Without warning the cliff gave way directly above the party. The fall of an estimated 100 tons of overhanging sandstone killed two women and a young girl on the spot,

Figure 3.7 A train passing along the sea wall at Coryton Cove, Dawlish, leaves Kennaway Tunnel heading south towards the northern portal of Coryton Tunnel. It was near this location, beneath the cliffs through which Kennaway Tunnel is cut, that the fatal accident of August 1885 took place (photo: R. W. Duck).

whilst the rest of the group were more or less seriously injured: 'A hospitable home has been darkened with lifelong regrets, and even Dawlish has been frightened.'[75] The three bodies were discovered in a frightfully mangled condition. These were subsequently identified by Lady Graves-Sawle as her ladyship's granddaughter, Violet Mary Watson, aged nine, Elizabeth Keen, Miss Watson's governess and Elizabeth Radford, companion to Lady Graves-Sawle. Though some were severely injured, the remainder of the party miraculously escaped death.[76] In its own inimitable extravagant style, the front page of *The Illustrated Police News* depicted the scene as the cliff collapsed (Figure 3.8).[77] The following Tuesday, another rock fall took place at the same spot.[78]

Figure 3.8 'Fatal accident thro' the fall of earth from the rock at Dawlish' (from the *Illustrated Police News*, 12 September 1885).

At the inquest, a verdict of 'accidental death' was recorded.[79] However, it was described as an accident waiting to happen in 'A death trap at the seaside'.[80] Furthermore, one witness noted that it had occurred within a few minutes of the passing of a down train.[81] As is so often the case at the coast, there was ambiguity over ownership of the beach and whose responsibility it should have been to warn people of the potential dangers. *The Times* summed up the situation:

> The railway people take laudable pains to attract people from far and near, and are successful; but they do not hold themselves answerable for their safety. The local Board claims the right to keep order in the multitude assembled there every morning in the summer, but has nothing to do with the soil they tread on, or the loose and fissured cliff, washed by the waves below, and trembling with the passing trains. The superintendent of the bathing machines evidently did not think it his business to warn people not to sit under the cliff, arguing that there could not be more danger in sitting quietly under it than walking under it, which his customers had to do in large bodies every morning.[82]

Furthermore, *The Times* advised those families accustomed to the comparative security of rural or town life, 'not to trust themselves to the coast where land and water, rushing trains and angry breakers, railway companies and local Boards, still contend for the mastery.'[83] Certainly, no-one associated with this terrible tragedy came away from the inquest covered with glory.

The down platform

The railway at Muchalls in Kincardineshire has surely seen more than its fair share of disasters over the years. Allegedly visited and much admired by the omnipresent Charles Dickens in its Victorian heyday, this sleepy cliff-top village near Stonehaven perches some 60–70 metres above the North Sea. Its station, however, was closed over 60 years ago and little trace of its former existence remains. Today, express trains speed through en route to or from Aberdeen offering sharp-eyed passengers but a fleeting glimpse of a curious memorial to World War I set into the seaward bank side of the cutting, close to the site of the former station. A mosaic of tiles depicts a royal crown and the word 'PEACE' above with the date '1919' below, set into a concrete shield. In the latter half of the nineteenth and the early twentieth centuries, however, this spot was far from peaceful, with a surprisingly large number of accidents, injuries and fatalities taking place on the line through Muchalls village.

Early in the morning of 27 September 1851, when the route was just two years of age and must still have been a novelty, the body of an elderly woman was found on the tracks. Boarding unconventionally at Muchalls, she had apparently attempted to travel northwards to Aberdeen by clinging onto the outside of a train carriage so as to avoid paying the fare. Sadly, but not surprisingly, she had fallen to her death. As if tragically personifying the Aberdonian stereotype, *The Aberdeen Journal* noted that, 'She is believed to have been somewhat silly'.[84] Further tragedy brought in the New Year of 1865; on the evening of 5 January two men walking along the line were run down by a train close to the station, one was killed, the other very seriously injured.[85] Two years later, in the early hours of Sunday, 13 January 1867, three men were killed in a collision between two pilot engines during snow-clearing operations.[86] On the morning of 23 March 1879, the body of a ploughman was found lying across the rails, having been struck and killed by the previous night mail.[87] Ten years later, on the afternoon of 9 March 1889, a passenger train from Aberdeen to Perth was approaching Muchalls station at about 60 miles per hour when a horse box coupled next to the engine left the rails. The remainder of the train, comprising three carriages, three brake vans and another horse box all followed, causing the line to be completely blocked in both directions. Remarkably no-one was injured.[88]

On the evening of 31 July 1891, a four-year-old child fell from a train as it passed through Muchalls station, the door at which he was standing suddenly opened while he was leaning against it. Although very severely injured, remarkably he escaped death.[89] In the early morning of 21 September 1897, a large log protruding from a heavily laden goods train bound for Aberdeen was the catalyst for yet another derailment, on this occasion on a spectacular scale. The brick signal box at Muchalls was virtually demolished as vans and wagons mounted the platform tearing up masonry and shedding their assorted loads; the line was completely blocked by debris and a wooden footbridge was destroyed. The aftermath at Muchalls station was described as one of 'hopeless confusion'[90] and this is borne out by a grainy photograph of the disaster that records the scene.[91] The following year, yet another dark January evening accident took place. On the morning of 13 January 1898, the body of a young man was discovered on the beach directly beneath the station. It transpired that two days earlier he had set off on foot to catch an evening train to Aberdeen. Making his way to the station in the darkness via the rough cliff-top path, he had presumably slipped and fallen to his death.[92] A more peaceful, accident-free period ensued for three decades until the afternoon of 12 June 1929, when a middle-aged man sustained severe injuries when he fell from a northbound train passing through the village.[93]

As if this catalogue of carnage and calamity was not sufficient, a remarkable, indeed unique occurrence took place the following year at Muchalls. The beginning of January 1930 was characterised by severe gales and heavy rain across much of Britain, leading to widespread flooding and landslides.[94] At Muchalls station a huge section of the eastern side platform was carried away in a large cliff fall.[95] Fortunately, the platform was deserted at the time and there were no injuries. As might have been surmised from the January accident of 1898, the station had been built right on the edge of the cliff with the seaward platform only a matter of feet from the lip. The bedrock in this area comprises metamorphosed sandstones known as psammites – flaggy, mica-rich and complexly folded metamorphic rocks – of the Dalradian Supergroup of late Precambrian age.[96] The ancient crystalline formations have been carved by the sea into spectacular wave-eroded landforms – caves, natural arches stacks and reefs – in an area known evocatively as Grim Brigs. Standing high and proud, the Old Man of Muchalls is the local answer to Orkney's the Old Man of Hoy. Little wonder that this not well known but very spectacular part of the Scottish coast was used as a backdrop by Franco Zeffirelli for his 1990 version of *Hamlet*. Here on the cliff edge at Muchalls a dummy set of ramparts was built for ghost scenes on battlements.[97] The cliff line itself is highly irregular and embayed; the sea has exploited any zones of weakness in the rocks such as softer layers, faults and joints, whilst more resistant rock masses stand proud as headlands. The bedrock is covered with a veneer of superficial deposits, laid down by ice during the Pleistocene glaciation of the area and likely re-worked subsequently by flowing waters and slope wash. This ill-sorted mixture of mainly sand and gravel-sized particles sits on top of a very irregular bedrock surface so that its thickness varies considerably, even over short distances laterally. The station platform that collapsed as a result of the heavy rains was founded directly onto the unconsolidated drift and to seaward of it is a deeply embayed inlet, V-shape in plan, known as a geo. At the time of the collapse it was reportedly known as the Strathbarrel Gully,[98] though this term appears to have fallen into disuse in favour of the perhaps more apposite Grim Haven, which includes the offshore continuation of the feature. It was an alarming incident:

The east platform at its north end used to be within a foot or two of the edge, but in the subsidence a large section of the platform has disappeared over the cliff. The displacement is estimated to be about 300 tons of material. Most of this, chiefly sandy gravel, fell into the sea, and has now been washed away, but many tons lodged on a ledge which juts out some distance at the foot of the cliff. The edge of the cliff, a 3ft. slice of the 6ft. platform, and a heavy sleeper fence which separated the two have gone, but the most alarming feature is that the outer rail of the main line is now only about 6ft. from the soft sandy edge of the precipice.[99]

Thus the southbound platform sat in a precarious location, directly above a dangerous piece of cliff. From a vantage point to landward there are signs of recent falls of unconsolidated materials down slope to the back of the beach (Figure 3.9). It is perhaps surprising that the

Figure 3.9 The cliffs at Muchalls looking northwards. The railway station platform that collapsed in January 1930 was located on the edge of the cliff in the foreground, in the left of the field of view, from which recently slipped debris can be seen accumulating on the cobble beach below (photo: R. W. Duck).

platform remained in place for so long at a spot noted for its landslides and cliff collapses, typically during the winter months.[100] Needless to say, at Muchalls, this main railway artery carries its heavy traffic in a perhaps imperceptibly but ever increasing position of coastal vulnerability. As our climate changes, this cliff-top section of line is a far from isolated instance of exposure to wind, wave and weather, as will be explored in Chapter 6.

Notes

1. Cruden, D. M. (1991), 'A simple definition of a landslide', *Bulletin of Engineering Geology and the Environment*, 43, 27–29.
2. 'Mr Clement Scott dog bitten', *Reynolds's Newspaper*, 17 July 1898.
3. 'Landslip at Cromer', *Reynolds's Newspaper*, 17 July 1898.
4. 'Evils we might avoid', *Reynolds's Newspaper*, 17 July 1898.
5. Emelijanow, V. (2004), Scott, Clement William (1841–1904), *Oxford Dictionary of National Biography*, Oxford: Oxford University Press, available at http://www.oxforddnb.com/view/article/35982 (last accessed 26 April 2014).
6. Cleveland, D. (1975), 'Poppyland', *The Lady*, 5 June, 1002–1003.
   Stibbons, P. and Cleveland, D. (2001), *Poppyland: Strands of Norfolk History* (4th Edition), Cromer: Poppyland Publishing, 29 pp.
7. 'Cromer', *The Morning Post*, 23 January 1825.
   'The ravages of the sea', *Manchester Times*, 9 November 1900.
   Ward, E. M. (1922), *English Coastal Evolution*, London: Methuen & Co. Ltd, 262 pp.
   Hutchinson, J. N. (1976), Coastal landslides in cliffs of Pleistocene age between Cromer and Overstrand, Norfolk, England, in Janbu, N., Jorstad, F. and Kjaernsli (eds), *Laurits Bjerrum Memorial Volume – Contributions to Soil Mechanics*, 155–182.
   Pennick, N. (1987), *Lost Lands and Sunken Cities*.
   Duck, R. W. (2011), *This Shrinking Land*.
8. Sea Wall Defences Including Promenade and Cliff Retaining Walls from Opposite the Bottom of Melbourne, Cromer, British Listed Buildings, available at http://www.britishlistedbuildings.co.uk/en-490171-sea-wall-defences-including-promenade-an (last accessed 26 April 2014).
9. Stibbons, P. and Cleveland, D. (2001), *Poppyland*.
10. 'Holiday traffic by rail and sea: Points for the tourist', *The Pall Mall Gazette*, 27 July 1898.
11. Scott, C. (1890), *Blossom Land and Fallen Leaves*, London: Hutchinson and Co., 322 pp.
12. The Idler's Club (1893), *The Idler*, 2, 218–232.
13. 'The destruction of Poppyland', *The Pall Mall Gazette*, 25 August 1891.
14. 'The destruction of Poppyland', *The Pall Mall Gazette*, 25 August 1891.
15. 'The destruction of Poppyland', *The Pall Mall Gazette*, 25 August 1891.
16. 'The destruction of Poppyland', *The Pall Mall Gazette*, 25 August 1891.
17. 'In defence of Poppyland', *The Pall Mall Gazette*, 31 August 1891.
18. 'In defence of Poppyland', *The Pall Mall Gazette*, 31 August 1891.

19. 'The ruin of the East Coast' (1898), *The Saturday Review*, 86, 233–234.
20. Stibbons, P. and Cleveland, D. (2001), *Poppyland*.
21. Stibbons, P. and Cleveland, D. (2001), *Poppyland*.
22. 'The ruin of the East Coast' (1898), *The Saturday Review*, 86, 233–234.
23. Stibbons, P. and Cleveland, D. (2001), *Poppyland*.
24. Plantlife, *Red Alert: Britain's Iconic Property Under Threat*, 29 November 2013, available at http://www.plantlife.org.uk/about_us/news_press/red_alert_ britains_iconic_poppy_under_threat (last accessed 26 April 2014).
25. 'London, Tuesday, May 10', *The Standard*, 10 May 1898.
26. 'The ruin of the East Coast' (1898), *The Saturday Review*, 86, 233–234.
27. 'Storm ravages on the coast. Slips on the Scarbro' and Whitby line.', *The Yorkshire Herald and the York Herald*, 2 March 1900.
28. Brandon, D. (2010), *Along the Yorkshire Coast: From the Tees to the Humber*, Stroud: The History Press, 159 pp.
29. 'The landslip at Whitby', *The Graphic*, 21 January 1871.
    'The Whitby landslip', *The Yorkshire Herald and The York Herald*, 24 February 1900.
    Duck, R. W. (2011), *This Shrinking Land*.
30. Quoted in: Brandon, D. (2010), *Along the Yorkshire Coast: From the Tees to the Humber*.
31. 'Landslip on the Scarbro' and Whitby railway', *The York Herald*, 8 August 1888.
    'Serious landslip on the Scarborough and Whitby railway', *The Leeds Mercury*, 11 August 1888.
    'The landslip on the Scarborough Whitby railway', *The York Herald*, 13 August 1888.
32. 'Serious landslip at Robin Hood's Bay. Railway communication stopped.', *The North-Eastern Daily Gazette*, 28 February 1900.
    'Summary of news', *The Yorkshire Herald and the York Herald*, 2 March 1900.
    'Great landslip on the Yorkshire coast', *Leicester Chronicle and Leicestershire Mercury*, 3 March 1900.
33. 'Storm ravages on the coast. Slips on the Scarbro' and Whitby line.' *The Yorkshire Herald and the York Herald*, 2 March 1900.
34. 'Landslides in Yorkshire', *The Scotsman*, December 26 1925.
35. 'A new watering-place for Yorkshire', *The Leeds Mercury*, 8 July 1896.
36. Duck, R. W. (2011), *This Shrinking Land*.
37. Sekon, G. A. (1895), *The History of the South-Eastern Railway*.
38. Dodd, J. C., Dougall, T. A., Clapp, J. P. and Jeffries, P. (2002), 'The role of arbuscular mycorrhizal fungi in plant community establishment at Samphire Hoe, Kent, UK – the reclamation platform created during the building of the Channel tunnel between France and the UK', *Biodiversity and Conservation*, 11, 39–58.
39. Anon. (1845), 'The coast line of the London and Dover Railway', *Penny Magazine of the Society for the Diffusion of Useful Knowledge*, 13, 257–259.
40. Anon. (1845), 'The coast line of the London and Dover Railway'.
41. Topley, W. (1893), 'The Sandgate landslip', *The Geographical Journal*, 1, 339–341.
    Muir Wood, A. M. (1955), 'Folkestone Warren landslips, investigations 1948–50', *Proceedings of the Institution of Civil Engineers*, 4, 410–428.

Anon. (1954), 'Coast erosion works in Folkestone Warren', *The Railway Magazine*, September 1954, 601–604.

Otter, R. A. (1994), *Civil Engineering Heritage: Southern England*, London: Thomas Telford Ltd, 295 pp.

British Geological Survey, Landslides at Folkestone Warren, Kent, available at http://www.bgs.ac.uk/landslides/folkestoneWarren.html (last accessed 26 April 2014).

42. 'Encroachment of the sea at Dover', *The Times*, 21 October 1841.
43. 'Encroachment of the sea at Dover', *The Times*, 21 October 1841.
44. 'Encroachment of the sea at Dover', *The Times*, 26 October 1841.
45. 'Destruction of the Round Down Cliff by gunpowder', *The Standard*, 27 January 1843.
46. 'Splendid engineering triumph', *The Era*, 29 January 1843.
47. 'South-Eastern Railway', *Lloyd's Weekly London Newspaper*, 29 January 1843.
48. Sekon, G. A. (1895), *The History of the South-Eastern Railway*.
49. 'South-Eastern Railway', *Lloyd's Weekly London Newspaper*, 29 January 1843.
50. 'The South-Eastern Railway', *The Times*, 8 February 1844.
51. 'The South-Eastern Railway', *The Times*, 8 February 1844.
52. 'The South-Eastern Railway', *The Times*, 8 February 1844.
53. 'Opening of the South-Eastern Railway to Dover', *The Morning Chronicle*, 7 February 1844.
    'Opening of the South-Eastern Railway from Folkestone to Dover', *The Standard*, 7 February 1844.
54. 'Block on the South-Eastern Railway', *The Lancaster Gazette and General Advertiser for Lancashire, Westmorland, and Yorkshire*, 17 January 1877.
55. 'Fatal occurrence on the South-Eastern Railway', *The Morning Post*, 16 January 1877.
56. 'The Dover and Folkestone Railway', *The Daily Gazette*, 8 March 1877.
    'The South-Eastern Railway', *The Morning Post*, 9 March 1877.
57. 'Big fall of cliff near Folkestone. Railway line blocked.', *The Times*, 21 December 1915.
    Anon. (1954), Coast erosion works in Folkestone Warren.
    Hutchinson, J. N., Bromhead, E. N. and Lupini, J. F. (1980), 'Additional observations on the Folkestone Warren landslides', *Quarterly Journal of Engineering Geology*, 13, 1–31.
58. Hutchinson, J. N., Bromhead, E. N. and Lupini, J. F. (1980), 'Additional observations on the Folkestone Warren landslides'.
59. 'Folkestone – Dover line reopening', *The Times*, 24 July 1919.
60. Anon. (1954), Coast erosion works in Folkestone Warren.
    Birch, G. and Warren, C. D. (2007), Landslide and chalk fall management on the Folkestone to Dover railway line, Kent, in McInnes, R., Jakeways, J., Fairbank, H. and Mathie, E (eds), *Landslides and Climate Change: Challenges and Solutions – Proceedings of the International Conference on Landslides and Climate Change, Ventnor, Isle of Wight, UK, 21–24 May 2007*, 295–304, Taylor and Francis.
    Bromhead, E. N. and Ibsen, M.-L. (2007), in McInnes, R., Jakeways, J., Fairbank, H. and Mathie, E (eds), *Landslides and Climate Change: Challenges and Solutions – Proceedings of the International Conference on Landslides and*

Climate Change, Ventnor, Isle of Wight, UK, 21–24 May 2007, 17–24, Taylor and Francis.

61. Hutchinson, J. N., Bromhead, E. N. and Lupini, J. F. (1980), 'Additional observations on the Folkestone Warren landslides'.

62. Mackie, S. J. (1883), *A Descriptive and Historical Account of Folkestone and its Neighbourhood with Gleanings from the Municipal Records, Reprinted from the Folkestone Express (2nd Edition)*, Folkestone: J. English, 352 pp.

63. Hutchinson, J. N., Bromhead, E. N. and Lupini, J. F. (1980), 'Additional observations on the Folkestone Warren landslides'.

64. Walker, A. and Musson, R. (2007), 'Kent rocked', *Geoscientist*, 17(6), 6.

65. Klose, C. D. (2007), 'Coastal land loss and gain as potential earthquake trigger mechanism in SCRs', *Eos, Transactions of the American Geophysical Union*, 88, Fall Meeting Supplement, Abstract T51D–0759.
    Klose, C. D. (2013), 'Mechanical and statistical evidence of the causality of human-made mass shifts on the Earth's upper crust and the occurrence of earthquakes', *Journal of Seismology*, 17, 109–135.

66. Great landslip on the South Devon Railway, *The Times*, 31 December 1852.

67. Great landslip on the South Devon Railway, *The Times*, 31 December 1852.

68. The South Devon Railway – Dawlish, *The Times*, 6 April 1853.

69. Kay, P. (1990), *Rails Along the Sea Wall*.

70. Lowe, E. J. (1865), 'History of the earthquake of 1863, October 6', *Proceedings of the British Meteorological Society*, 2, 55–99.
    Davison, C. (1927), 'The Hereford earthquake of 15th August, 1926', *Geological Magazine*, 64, 162–167.

71. 'The earthquake. To the Editor of The Times', *The Times*, 8 October 1863.

72. 'The earthquake of yesterday. To the Editor of The Times', *The Times*, 7 October 1863.

73. 'Terrible accident', *The North-Eastern Daily Gazette*, 29 August 1885.

74. 'Editorial', *The Times*, 2 September 1885.

75. 'Editorial'. *The Times*, 2 September 1885.

76. 'Shocking accident at Dawlish', *Trewman's Exeter Flying Post*, 2 September 1885.

77. 'Fatal accident thro' the fall of earth from the rock at Dawlish', *The Illustrated Police News, Law Courts and Weekly Record*, 12 September 1885.

78. 'Further fall of cliff at Dawlish', *The Standard*, 2 September 1885.

79. 'The fatal accident at Dawlish', *The Times*, 1 September 1885.

80. 'A death trap at the seaside', *Birmingham Daily Post*, 2 September 1885.

81. 'The fatal accident at Dawlish', *The Bristol Mercury and Daily Post*, 2 September 1885.

82. 'Editorial', *The Times*, 2 September 1885.

83. 'Editorial', *The Times*, 2 September 1885.

84. 'Melancholy accident', *The Aberdeen Journal*, 1 October 1851.

85. 'Fatal accident on the Scottish North-Eastern Railway', *Glasgow Herald*, 9 January 1865.

86. 'Distressing railway accident near Muchalls – three men killed', *The Aberdeen Journal*, 16 January 1867.
    Rich, F. H. (1867), Caledonian Railway, *Report of The Secretary of the Railway Department, Board of Trade*, 3–4.

87. 'Frightful death on the Caledonian Railway', *Dundee Courier and Argus*, 24 March 1879.
88. 'Serious railway accident at Muchalls', *Dundee Courier and Argus*, 11 March 1889.
89. 'Child falls from an express train', *Dundee Courier and Argus*, 1 August 1891.
90. 'Railway accident near Stonehaven', *Dundee Courier and Argus*, 22 September 1897.
91. Watt, B. H. (2005), *Old Newtonhill and Muchalls*, Catrine: Stenlake Publishing Ltd, 49 pp.
92. 'Fatal accident at Muchalls', *Aberdeen Weekly Journal*, 19 January 1898.
93. 'Fall from north train', *The Scotsman*, 13 June 1929.
94. 'Another gale', *The Times*, 3 January 1930.
95. 'Railway platform in cliff fall. Main line near the edge.', *The Times*, 7 January 1930.
96. Thomas, C. W. (1999), 'Late Caledonian calc-alkaline dykes from the east coast of Aberdeenshire', *Scottish Journal of Geology*, 35, 1–14.
97. WhereDidTheyFilmThat.co.uk, *Hamlet* (1990), available at http://www.wheredidtheyfilmthat.co.uk/film.php?film_id=112 (last accessed 26 April 2014).
98. 'Railway platform in cliff fall. Main line near the edge.', *The Times*, 7 January 1930.
99. 'Railway platform in cliff fall. Main line near the edge.', *The Times*, 7 January 1930.
100. 'Community councillors shocked by storm damage', *Mearns Leader*, 25 January 2013, available at http://www.mearnsleader.co.uk/news/local-headlines/community-councillors-shocked-by-storm-damage-1-2756206 (last accessed 26 April 2014).

4

Across Salt Marsh, Mudflat, Slob and Sleech

In these anxious times for the people of the United Kingdom it is becoming clearer every year that there is not enough room for them all to live comfortably at home. Unless industrial development keeps pace with the growth of the population, the congestion will become still worse.

H. M. Cadell, *Land Reclamation*
in the Forth Valley, 1929

Introduction

In late nineteenth-century Britain the sea was, quite simply, the enemy. *The Times* encapsulated this bluntly: 'it behoves every Englishman to protect his land as far as he can from a more powerful destroyer than human potentates.'[1] Furthermore, if what at that time was referred to as 'wasting' of the coast was indeed taking place, the ethos was very much, 'an eye for an eye'. If land was being lost to the sea by erosion at one location, it was our duty to our country to 'win' land back elsewhere to compensate.[2] Claiming land from the sea was also seen as vital to sustain Britain's increasing population. In the south east of England, for instance, there was:

> ... a vision, not only of dairy farms, but of roads, and villages, and church spires, and parish schools, and a permanent enlargement of the county of Essex – not by the mere reclamation of a fen, but by the positive embankment of the sea sands, shutting out the ocean and transforming the sandy floor into solid ground, fit to build and live upon.[3]

This industrial revolutionary 'vision' was typical of many low-lying coastal regions across the length and breadth of Britain – and railways were often inextricably bound up in it.

Henry Moubray Cadell was, amongst his many talents, a geologist who, for a brief period during the late nineteenth century, worked for the Geological Survey of Scotland. A polymath, he was a pioneer of experimental geology and has become famous amongst historians of earth science for his 'squeeze box', within which different coloured layers of sand could be compressed gradually to simulate the folding and fracturing of rocks within mountain belts.[4] Cadell was also a land-owner, mining consultant, industrialist, company chairman, patron of the arts, county councillor, Justice of the Peace, founder member of the Royal Scottish Geographical Society, frequent overseas travel-ler to distant parts, father of eight children, prolific writer of letters to *The Scotsman* and an accomplished author. An early proponent of renewable energy from tidal power, he was a passionate advocate for land reclamation, especially around the fringes of his native Firth of Forth.[5] The Cadell family seat was that of Grange in West Lothian on the southern side of the estuary between Bo'ness and Linlithgow. In 1913 he published his magnum opus, *The Story of the Forth*,[6] in which the progressive enclosure of inter-tidal salt marsh, (often denoted on eighteenth- and nineteenth-century maps as salt greens or saltings) mudflat, slob and sleech from its shores was chronicled. To Cadell's practical mind – like those of so many industrial revolutionists – such areas were simply 'waste lands'. Turning them over to fertile agricul-tural production or to the foundations for harbours and waterfront industrial developments, linked by railway arteries, as at Culross and Low Valleyfield (see Chapter 1), was a venerable cause. He saw improved agricultural productivity around the Forth, as in other areas of Britain, as a means of holding communities together:

A large rural population is a pillar of strength, moral as well as mate-rial, to every community, and anything that can be done to prevent the depopulation of our rural districts should be attempted, for *A bold peasantry, their country's pride, When once destroyed can never be supplied.*[7]

Cadell certainly had a point and, as such, he saw land claim as being foresighted and an enterprise for which those involved should be con-gratulated. It has already been mentioned (Chapter 1) that around one half of the area of inter-tidal lands of the Firth of Forth have now been destroyed by human endeavour. To Cadell this would, no doubt, have been a source of great satisfaction and pride but he would have seen it as unfinished business. Since his death in 1934, however, understanding, attitudes and perceptions have changed beyond all measure. We have reassessed our environmental values and no longer consider such areas as 'waste' but realise the enormous ecological importance and value of estuarine salt marsh, mudflat, slob and sleech as reservoirs of biodiversity. Not only do retaining embankments and walls result in modifications to natural tidal and sediment flow pathways and speeds, the areas of once natural, inter-tidal habitats that have been destroyed are no longer able to provide, among others, bird breeding or feeding sites, fish spawning or juvenile feeding grounds. So too, the natural capacity for estuaries to respond to natural coastal changes and rising sea levels has become constricted. In the Forth, a study of the impact of land claim on estuarine biota has shown that a simple consideration of area lost may be inadequate to describe its effects, and the consequences for animals living within any estuary affected in this way may differ considerably amongst themselves and from area to area.[8] Specifically within the Forth, the actions that Cadell so avidly supported have collectively removed 24% of the natural fish habitats, which equate with the removal of as much as 40% of their food supply. Furthermore, the impact of land claim on local bird populations varies according to species but the large scale loss of habitat has reduced the size of some shorebird populations which overwinter in the estuary.[9]

The Forth, however, is by no means the most affected by land claim of the major British estuaries. That dubious honour falls to the Tees in the north east of England. Although more land has been claimed from The Wash in East Anglia than any other estuary in Britain, the much smaller Tees Estuary has lost the greatest percentage of its inter-tidal area to the extent that it has been almost completely destroyed. In the 1850s the Tees was characterised by huge expanses of inter-tidal sand and mudflats fringed by salt marshes; only minor amounts of land had been claimed for agriculture. This, however, was soon to change. The revolutionary pace of growth of the iron and steel industry – fuelled by nearby plentiful finds of coal and iron ore – along with the need to build new docks and harbours, was so great in the latter half of the nineteenth century that the estuary's inter-tidal areas became destroyed on all flanks by successive and rather piecemeal episodes of land claim. These actually continued until as recently as the 1970s, by which time, an incredible 83%[10] of the inter-tidal area of the Tees Estuary had been claimed for the needs of industry, including now largely long-disused myriads of railway sidings, marshalling yards and quayside branches.

The Southwold Railway – christened with a flood

Railway construction across low-lying reclaimed land was certainly not without its problems. For instance, the Blyth Estuary in rural Suffolk has a long history of land claim and channel straightening to improve navigation.[11] To increase agricultural productivity, large tracts of salt marsh have been drained and enclosed since the sixteenth century, extending inland from Southwold at the coast to the tidal limit near Halesworth.[12] The isolation of the aspiring watering place of Southwold – itself no stranger to the ravages of the North Sea[13] – from the Great Eastern Railway system was rectified in 1879 with the opening of the Southwold Railway. This branch line from Halesworth, 14 kilometres in length, constructed without opposition from the local landowners,[14] was unusual owing to its narrow gauge of 3 feet.[15] The line was built largely on low embankments crossing previously reclaimed, low-lying marshlands but additional drainage of inter-tidal lands was required along with a swing-bridge spanning the artificially straightened River Blyth. It opened for public traffic on 24 September 1879 and had a maximum speed limit of 16 miles per hour:

Southwold was gaily decorated for the occasion, the Church bells rang a merry peal, and the day was observed as a general holiday by the towns-people, and in order to provide amusement for the holiday seekers a capital programme of athletic and other popular sports was arranged.[16]

Despite the jubilations at the eastern terminus, the unfolding events of the opening day of the Southwold Railway's operations did not bode well for the line's future. A 'well freighted' train ran in the morning from Southwold through to Halesworth, according to the timetable. On its return journey it was equally well filled with passengers, 'thus affording gratifying promise of a successful opening day in point of public patronage.'[17] That promise was short lived, however. Deteriorating weather conditions, coinciding with high tide, led to extensive flooding during the course of the afternoon. In consequence, 'hundreds of acres' of marshland around the railway formation became flooded. Moreover, the rising waters inundated the new railway along a stretch near to the village of Wenhaston. The extent of the flooding was such that the directors of the line took the decision to suspend the running of trains between Halesworth and Wenhaston until the waters had receded to a safe level. Trains were, however, run to suit the requirements of the public to and fro between Southwold and the temporary terminus and 'notwithstanding the threatening character of the weather during the afternoon, a comparatively large number of people were conveyed over the line.'[18] The 3-feet gauge of the Southwold Railway was unique in Britain but this line must surely be the only railway that became severed from its main artery due to flooding on its very first day of public operation. Needless to say, given the line's route, largely over land that was once part of the Blyth Estuary, this occurrence was not a one-off.

On two occasions during the winter of 1894 to 1895 high tides yet again coincided with river floods inundating large tracts of reclaimed salt marsh around Southwold and making two large breaches in the railway embankment, carrying away the ballast and leaving the single line suspended in the air.[16] Three years later a severe gale, coinciding with a spring tide, hit the eastern and southern coasts of England. Monday, 29 November 1897 was to become known as 'Black Monday' in East Anglia.[20] Embankments were overtopped for almost the entire length of the coast from Norfolk to Essex, inundating vast areas of agricultural land.[21] As well as flooding, there was widespread structural damage and on the night of 2 December the continuing storms claimed the lives of nine men aboard the volunteer life boat *Friend to all Nations*, which capsized off Margate.[22] The embankment of the railway between the bridge over the River Blyth and Southwold Common was a less well known victim of 'Black Monday'. It was carried away by the flood, indeed such was the water level that the town was cut off from all communications with the mainland:[23] 'The damage done at Southwold by the storm was so great as to induce apprehensions that the whole town would ultimately be washed away by the sea.'[24] The early afternoon train from Halesworth on 29 November could only pass for a short distance before being met by the rising flood waters. Passengers and goods had to be transferred to Southwold in boats.[25] Frequently flooded, damaged and repaired, the line remained viable for five decades and survived until 1929.

Rails along the Eden

Whilst the Tees is an extreme case, almost no estuary, large or small, in lowland Britain has escaped the effects of land claim. Away from large scale industrial schemes of which railways are or once were an integral part, the fringes of far less developed water bodies have also been impacted – albeit in many cases insidiously – by the construction of railway embankments. The Eden Estuary of Fife– small in size and essentially rural in character[26] – is just one of the many that bears the scars of railway construction. However, it is the architect of the branch railway that once traversed its gently rolling shores, coupled with his parsimonious methods of construction, that add extra seasoning to the Eden's story.

On 30 March 2010, much of Scotland was taken by surprise, besieged by a combination of snow, rain, high tides and gale-force winds causing widespread havoc on land and sea. Many coastal communities on the east coast were battered particularly severely; there was extensive damage to boats, harbours, promenades, sea walls and houses in Fife, the Firth of Forth and East Lothian.[27] One victim of this storm that did not reach the national press and television coverage was the St Andrews railway or, to be more precise, what remains of its formation today. The old embankment that once carried the line along the southern shore of the Eden Estuary was breached in two places by wave attack. Held back for over 160 years, these gaps, eroded through the barrier by storm waves, permitted the sea to once again flood fields that would have naturally been inundated during high tides. This could have been an opportunity for 'managed retreat', to accommodate the increased volumes of water associated with sea level rise. However, the embankment was repaired by the landowner – at a cost of £75,000.[28]

The last train to travel along the short branch railway between St Andrews and Leuchars Junction did so in 1969. However, this link between the 'Home of Golf' and the outside world has left behind or at least contributed to an enduring coastal legacy. The line, around 7 kilometres in length, opened in 1852, several years after the peak of railway mania in the mid-1840s.[29] The place name 'Leuchars' is derived from *luachair*, meaning reeds or rushes in Gaelic and is a reminder of the ancient natural, character of the Eden Estuary's hinterland. Today the station is located on the East Coast Main Line but in the 1850s it was part of the Edinburgh, Perth and Dundee Railway Company's trans-Fife network. When a special train carrying the great and the good of the county, along with the directors of the St Andrews Railway Company, inaugurated the route at mid-day on Tuesday, 29 June 1852, the time occupied in travelling from Leuchars to the St Andrews station was between 13 and 14 minutes. Indeed at this gentle pace the motion of the train's carriages was described as 'remarkably smooth'.[30] The proximity of the line to the southern shore was remarked upon, as the train travelled eastwards from Leuchars, '. . . being high water at the time, the full tide of the Eden [was] close on the left and the green corn fields on the right . . . '[31] As was typical of such an occasion, the day was rounded off with an all-male dinner in the St Andrews Town Hall, which was 'tastefully decorated with evergreens. Seventy-three gentlemen were present; and the dinner and speechifying went off with much éclat.'[32]

The St Andrews Railway was the first of the so-called 'Cheap Railway Movement' in Scotland. Its engineer was a young Thomas Bouch. Not yet 30 years of age and newly established as an independent consultant after leaving his post as the Edinburgh, Perth and Dundee Company's Resident Engineer, this small branch line was to be his first commission.[33] The part that Bouch was to play in the infamous Tay Bridge disaster of 1879 and the notoriety he would gain thereafter could never have been foretold at this juncture. However, the economies that Bouch made in the construction of the St Andrews line were perhaps a portent of his future *modus operandi* and his ultimate personal downfall.

Except for a short distance where it crossed over a northward projecting spit known as Coble Shore, along much of its course the line was to hug the low-lying, southern shore of the Eden Estuary. This small, shallow water body, to the north of St Andrews, occupies the lower, coastal reaches of the River Eden and drains gently undulating, rich and fertile farm lands. Well-drained fields sloping down to the estuary gave way naturally to salt marsh, inundated on high spring tides, and then to inter-tidal mud and sand flats. Centuries previously and prior to much improved drainage in the area, the marshes around the Eden Estuary had been even more extensive and 'must have been well calculated for enjoying the sport of hawking.'[34] Building a railway over the relatively flat coastal lands offered few difficulties, though an embankment along much of the route where it skirted the estuary would be required. Bouch designed the line at a fee of £100 per mile of line, at a time when, by his own estimate, engineering fees in Scotland were averaging £500 per mile. Beginning with the St Andrews line, he was to establish himself by catering for enthusiastic local patriots, tailoring his lines to the capital funding available and, in this specific instance, raised largely from local subscribers.[35]

At the zenith of railway mania, many companies failed to estimate accurately the cost they would have to pay for the land required to build their lines and fell foul of the 'avaricious cupidity' of landowners, who attempted to squeeze every last drop of blood from railway companies. If, however, local landowners were sympathetic to or even involved in a particular scheme, then the costs of land acquisition were far more likely to be within budget. Such was the case with the St Andrews line; the landowners affected were generally well-disposed to its construction. Of the six or so principal landowners, one was the St Andrews Town Council, one the Bank of Scotland and two became directors of the newly formed company.[36] Most of the land needed was thus acquired amicably. Indeed Bouch was able to obtain the land he needed for the route for a little over half of his original estimate of £10,200;[37] surely an added fillip to his passion for economy. The landowners whose fields sloped gently northwards, down to the southern fringes of the Eden Estuary were, in fact, more than simply well-disposed. More to the point, they were exceedingly shrewd and, furthermore, far-sighted.

Three principal economies were deployed by Bouch on the St Andrews route. First, the use of lighter weight rails than normal – 60lbs per yard section compared with the customary 75lbs per yard in use at the time – was specified. Moreover, these were obtained secondhand from the Edinburgh, Perth and Dundee Railway. Secondly, the sleepers – untreated with any preservative such as tar – were to be spaced at 4-feet intervals instead of the customary 3 feet and the light cast iron chairs were to be secured to the sleepers with hardwood pegs, or treenails, instead of spikes. Thirdly, the line's two bridges were to be of timber construction (again untreated) rather than of masonry.[38] As such, this was the pioneer of the 'light railway', although the term was not used formally until the Regulation of Railways Act of 1868,[39] along which only light engines were to be used and run at moderate speeds of no more than that of the inaugural train. There was, however, a fourth and very important component to Bouch's economy in building the St Andrews line. He made use of two sections of previously constructed embankment along the southern shore of the Eden Estuary, thereby reducing the quantity of construction materials required. By joining these sections together and by widening them so as to carry the single track line, Bouch was able to build the railway embankment cheaply and in such a way that, along much of the southern shore of the estuary, it effectively formed the sea wall. In places, the line of the embankment was advanced or diverted seawards so that even more land was claimed from the estuary for use by the local farmers.[40]

In the first two decades of the nineteenth century, attempts had been made to claim what would have been regarded as waste lands from the shores of the estuary to increase farming activity: 'About 55 acres of ground contiguous to the estuary of the Eden, and which overflowed part of it every tide, and the remainder at spring-tides, have at two different periods, been secured by embankments, and are now under cultivation.'[41] The first embankment was constructed around 1810 'on the lands of Strathtyrum'.[42] It was some 300 yards in length, and was 'executed at great expense.' The second construction was made around 1820 along the shore of Kincaple and 'at much less expense than that of Strathtyrum'.[43] The dike [sic] was said to have been:

> 1,100 yards in length, 9 feet in height, 30 feet broad at the base, and 2½ broad at the top, with a slope of 21 feet on each side. It was constructed of pure sea sand brought from the sea beach, which is entirely covered with a coating of thick turf; and the side next the sea is faced with stonework about 12 feet up the slope. For the exit of water from within, there are two iron pipes, each 14 inches in diameter, which form sluice drains, having brass valves fitted into them, which shut or open as the tide advances or recedes. The whole expense of this structure was £1,250 sterling.[44]

As a consequence, this retaining structure claimed 30 acres of land from the inter-tidal zone on the southern shore of the Eden Estuary,[45] thereby expanding the area of the Kincaple estate. It is not surprising to find that the sand required to build the bulk of this embankment was won from local sources (see Chapter 5), the nearby beaches of the outer part of the Eden, nor that it was known as the 'Salt-Grass dyke', a name which continued to be used in local guide books until the end of the nineteenth century.[46] It has been reported that, at one time, a plan was under consideration to 'secure' a much larger extent of the estuary, but those landowners interested in the undertaking doubted whether the expense would be justified so the scheme was abandoned.[47] In effect, Bouch's embankment finished off the job for them and, importantly, at no cost to themselves. Moreover, there is little doubt that Bouch did likewise source his materials locally to enlarge the two sections and to join them into a continuous embankment to carry the St Andrews Railway.

It is hardly surprising that Bouch's cost-cutting measures were soon to lead to the need for expensive remedial works. Almost from the beginning, the track work gave trouble and, in particular, the rails began to splay on curves. The untreated timberwork of the bridges deteriorated rapidly and began, quite literally, to fall apart.[48] The durability of the embankment was also brought into question as, just four months after the line opened, it became badly damaged in a terrific gale that took place on 26 October 1852. This caused widespread damage, losses of life and destruction to property and vessels in eastern Scotland and north east England. At Granton, for instance, a large part of one of the sheds erected on the pier for the storage of goods was washed away, along with a considerable portion of its contents, including a number of empty casks.[49] A fortnight or so after the storm had abated, a correspondent to the *Fifeshire Journal* wrote:

SIR – A dozen lines, if you please, of the *Journal* to draw the attention of the Directors and the Public to what appears, if not positively dangerous, at least very dangerous like, and not at all agreeable to a man of my nerve. I refer to the unprotected state of the railway embankment where it skirts the estuary of the Eden near to the Kincaple siding. During the late storm, the waves seem to have been breaking over the very rails, or at all events, have undermined the sandy embankment close under the sleepers – giving two and separate portions of the line anything but a safe appearance. A fortnight has now nearly elapsed since the storm abated, and nothing has yet been done by the Directors to fill up these breaches and render the embankment more substantial and safe. At whose door the fault lies, your humble servant knows not; but this he knows – that should an accident happen from the rails breaking down at these spots, the Directors will then learn to their cost whose duty it is not to break the heads and limbs of the Public. A hundred pounds or so would raise a substantial barrier, but ten or even twenty hundred may not cover the damage resulting from a culpable breakdown.[50]

On 16 May 1864 an accident took place on the line close to Guard's Bridge, then the port of Fife's County Town of Cupar and today more commonly known as Guardbridge. This was to be very revealing as to the state of the line some 12 years after its opening. A train consisting of a tender in front, an engine, one guard's van, one second, one first, and one third class carriage, coupled in that order had left Leuchars Station at 3.30 p.m.:

It left Guard's Bridge, about 3.39 p.m., being about five minutes late at both stations. About 300 yards beyond Guard's Bridge, the tender left the rails on the outside of a 30 chain curve, the engine followed, and the leading wheels of the guard's van next to the engine, also got off the rails.[51]

None of the passengers on board was injured and the train was reported to have been travelling at an appropriate speed for the line of about 10–12 miles per hour. Even those not conversant with contemporary railway terminology can be in any doubt as to the state of the line at the time and the cause of this derailment. Following the official inquiry into the accident, The Secretary of the Railway Department, Board of Trade, Captain F. H. Rich reported thus:

> The line itself is of very light and poor construction. The rails are single headed, in lengths of 16 feet, and weighed 65lbs. per yard lineal when laid down about 12 years since. They are considerably worn. The joint chairs weigh 22lbs. and the intermediate chairs 15lbs. each. They are weak, too narrow in the throat to admit a good key, and reported to be continually breaking. They were originally fastened to the sleepers with treenails, but a great number of spikes have now been substituted; the joints, in some cases are still fastened with treenails only.[52]

Captain Rich went on to state in no uncertain terms that: 'I consider that the whole permanent way wants renewing . . . The Company have got some new rails on the ground for relaying the line, and the sooner it is done the better.' The final, short, sharp sentence of the Captain's report into the accident was unequivocally damning: 'The accident of the 16th ult. [i.e. 16 May 1864] was caused by the bad state of the St Andrew's Railway',[53] the consequence of Thomas Bouch's cost-cutting.

Though the embankment *per se* was not implicated in the 1864 accident, the section of line alongside the Eden near Guardbridge continued to give concern because of storm damage and eventually the company was forced to act after a particularly severe winter storm. On Hogmanay of 1876, a major gale struck many areas of Britain from Tayside to the English Channel coast.[54] Heavy, incessant rain gave way to snow and there was widespread flooding and associated destruction. The severe weather continued to herald in the New Year of 1877 with a vengeance; for instance, the pier and promenade at Hastings were severely damaged and 2,000 tons of rock fell from the nearby cliffs. The sea wall at Brighton was breached and numerous vessels around the coast were smashed and swept away along southern and eastern shores. A portion of the railway from St Helier to St Aubin on Jersey was washed away; in Dover the massive upper promenade of Admiralty Pier was destroyed, the damage estimated at over £50,000, and the South-Eastern Railway Station was inundated.[55] In Sussex the railway between Pevensey, near Eastbourne, and Bexhill (see Chapter 5) was washed away and the whole of the land adjoining it was submerged; so too 150 yards of Eastbourne pier was destroyed in the gale.[56] The stories of destruction and devastation go on. Across Fife, 'Not for a period of upwards of twenty years has such a storm as the present prevailed throughout the county.'[57] At St Andrews, as elsewhere, the sea ran very high causing widespread flooding: 'On all the low lying ground there was an enormous quantity of water and the repeated storms have so endangered the St Andrews railway that at portions trains have to be run dead slow.'[58] It was reported that the railway between Leuchars and St Andrews had suffered more from this storm that it had done on any previous occasion. The strong gale from the east sent the water inland and the ballast had become much like mortar. To prevent overtopping by waves and the erosion of ballast, the line was raised above its former level nearly 2 feet. Yet again and perhaps not surprisingly, the new ballast 'was obtained from the banks along the sides of the line . . .'[59]

The 'Home of Golf' got its railways but not a promenade

A few retired elderly gentlemen amusing themselves at golf, or the education of a too limited number of youths within her precincts, do not comprise all that St Andrews is capable of.

George Bruce, Footnote in *Destiny and Other Poems*, 1876

In 1887 St Andrews ceased to be a railway terminus. The opening of the Anstruther and St Andrews Railway along with a new, centrally located station, permitted travellers from all of the small towns and villages around the Fife coastal peninsula, to the east, the south and the west, direct access to the town.[60] That same year, the sturdy successor to Thomas Bouch's ill-fated Tay Bridge was opened; thus swifter opportunities for passengers to visit St Andrews from Dundee and points north of the Tay became renewed. As a town with little or no manufacturing industry, dealing in people – tourists, golfers, university dons and their students – rather than goods, was vital to its prosperity. What St Andrews lacked was a promenade. Unlike many watering places around the country, the coming of the railway had not been accompanied by an esplanade for taking the sea air on foot or by carriage. That was to change, at least in part, in the late 1880s when two prominent citizens proposed rival schemes. The architect John Milne proposed the construction of a carriage esplanade bounded by a concrete sea wall, enclosing some 12.5 acres of land claimed from the sea.[61] After much wrangling, in 1893 Milne's rival on the Town Council, George Bruce, began the construction – at his own expense – of what is today known as the Bruce Embankment. As well as being a local politician, Bruce was a keen golfer and prolific poet.[62] His scheme involved the closing off of a small bay to create not a promenade as such but a roughly square shape of land immediately to the east of the mouth of golf's most famous watercourse, the Swilcan Burn, which forms a hazard on the first and eighteenth holes of the Old Course. Bruce achieved this by placing four obsolete fishing boats weighed down with stones across the mouth of the rocky bay to form a dam. Great quantities of the town's refuse were then shifted there by horse and cart and dumped to back-fill the bay.[63] Indeed, late nineteenth-century ceramic storage jars and glass bottles were retrieved from a recent excavation through the site.[64] However, as Milne had predicted, a concrete sea wall was needed to secure the infill of the embankment;[65] in a very severe gale in mid-October 1898, said to be the most severe of the past 30 years, thousands of tons of debris were washed away from the landfill, making 'a considerable inroad upon the links'.[66] Today the Bruce Embankment is given over largely to car parking and a putting green; the first tee of the world-famous Old Course is now around 100 metres from the sea compared with the few metres that it was originally. Bruce's scheme received mixed reaction. The local 'tourist' guide to the town proclaimed that:

> For the inception and successful carrying out of this scheme, the citizens are indebted to the devotion and zeal of Mr George Bruce. Since his bulwark was constructed in the autumn of 1893, a large piece of land has been reclaimed from the sea, and within a few years a substantial addition will have been made to our recreation ground. Already the rock known as the Big Doo-Craig can be reached at all states of the tide.[67]

By contrast, the author of the local 'natural history' guide for ramblers took the stance that: 'The general effect does not commend itself to all. The concrete wall is a feature difficult to tolerate'.[68] To the north of the Bruce Embankment, the fringing dunes of the West Sands extend unbroken for almost 3 kilometres to the spit, known as Out Head, at the mouth of the Eden Estuary. In Bruce's day, this extensive 'waste land' of wind-blown sand must have seemed the ideal location to extend an esplanade. Had this taken place, it would have been not only an ecological travesty; the retaining wall would have starved the natural longshore sand transport to the north, thereby accelerating erosion of the golf courses of the northern parts of the St Andrews links. For once, a coastline was not violated like so many others, as echoed by the natural historian who had the last word on the matter:

A preposterous proposal has occasionally been mooted to construct a promenade, after the pattern familiar to *habitués* of the gay English and other seaside resorts, as an extension of the Bruce Embankment along the West Sands. It is hardly conceivable that any lover of St. Andrews, whether native or not, would support a proposal of this kind.[69]

The Bruce Embankment withstood the force of the waves for over 100 years but the structure was beginning to show its age. A £1.5 million improvement scheme, including reconstruction and strengthening of the sea wall, was completed in 2005.[70]

The Eden embankment (Figure 4.1) has, with periodic repairs,[71] such as those necessitated by the 2010 storm, survived intact to the present

Figure 4.1 The breach in the former railway embankment on the south side of the Eden Estuary on 1 April 2010, resulting from the storm of two days earlier. The extent of inundation and erosion of the adjacent fields provides a measure of the area of naturally inter-tidal land and salt marsh that has been lost owing to embankment construction (image courtesy of Professor Robert M.M. Crawford, University of St Andrews).

day. Its longevity is such that the salt marsh that it cut off from tidal influence and gave over to increased agricultural activity, is largely long forgotten. It is rather ironic that the *Fife Coast and Countryside Trust*, part of the European Union SUSCOD or SUStainable COastal Development in practice[72] project, is considering the design of a managed realignment scheme in the Eden Estuary. Their study involves the potential effects of removing part of the old railway embankment, which might help to increase inter-tidal habitat. The Trust has also suggested that such a breach might also bring added economic benefits to local landowners, if they accept the partial inundation of their agricultural lands in exchange for improved public access and new recreational and educational facilities. In light of the response to the March 2010 event, it would appear that the latter argument still remains unconvincing. Should the railway to St Andrews ever be re-opened – and there is a strong head of steam behind such a proposal[73] – it will follow an inland route from the main line to the town.[74]

Contrary to popular belief, the St Andrews railway was not a direct victim of Dr Richard Beeching's axe. His infamous 1963 'report' merely advocated the closure of the one intermediate station at Guardbridge.[75] The branch from Leuchars survived these swingeing cuts; but the British Railways Board finally culled the line six years later. Beeching was, however, firmly responsible for the closure of another small branch constructed in a similar coastal setting at the opposite end of the country – between Wivenhoe and Brightlingsea in Essex.

A lonely way in Essex

With its numerous meandering creeks, estuaries and inlets, the coastline of Essex is complex; furthermore it is around 600 kilometres in length. Between the River Stour in the north and the River Thames in the south, this is a land of marshes, mudflats and saltings, vulnerable to storms and to relative sea level rise. It is little surprise that the county has a long history of coastal defences, comprising sea walls and embankments. Vast areas of marshlands have been reclaimed over the centuries and defended from the natural encroachment of the sea. Nearly completely encircled by a combination of muddy creeks and the estuary of the River Colne, the town of Brightlingsea is almost located on an island. It is a Limb of the Head Cinque Port of Sandwich and, as such, is the only port outside Kent and Sussex with a connection to the ancient Confederation of the Cinque Ports[76] In *The Domesday Book* the place was called *Bricrtiseseia*, or 'island of a man called Beorhtric or Beorhtling'.[77] Indeed it has been isolated in the past as this low-lying part of Britain has succumbed to coastal flooding on numerous occasions over the centuries, most notoriously in 1953.

Dr Beeching's infamous axe fell here in 1964,[78] closing the short branch line from Wivenhoe that had opened up Brightlingsea, famous for its oysters, to the rest of Britain a century earlier on 17 April 1866.[79] This single line on a narrow track bed, worked by the Great Eastern Railway Company, hugged the eastern bank of the Colne Estuary for much of its length, crossing Alresford Creek by means of a large iron swing bridge, nearly 150 metres long. The line was not an easy one to build, necessitating 'engineering works of unusual difficulty and magnitude'.[80] A new embankment had to be constructed to carry the rails over almost all of the 8 kilometres to the junction with the Great Eastern's main line at Wivenhoe. Crossing creeks and deserted marshland, in places this was built adjacent to a pre-existing sea wall, which dated back to the sixteenth century, on its landward side. Along other stretches, however, it had to be built to seawards across the inter-tidal mud flats; some creeks were blocked off, others were diverted through the embankment via culverts.[81] In this way, the railway formation created a hard engineered defence along the eastern edge of the Colne Estuary. In addition, vast areas of salt marsh were reclaimed and cut off from tidal inundation. The coming of the railway put an end to several popular marshland walking routes, as the new line severed these. Reports of prosecutions for crossing the line were frequent in the local press.[82] The primary history of this short branch is, however, one of a continuous battle with the sea; its location was one of the most susceptible to breach or flooding in Britain. Indeed, the sea won on numerous occasions and the frequency with which the line was damaged is a measure of its vulnerability. Immediately prior to its opening, it was reported that:

The least noticed, although the most important portion of the whole line, is the river wall, protecting the railway. It is formed of the clay and mud excavated to a depth of from 2 feet 6 inches to 3 feet, the slope next the water being extremely flat, and the bank itself standing considerably above the railway inside it.[83]

The line had only been in operation for eight years when the first floods inundated parts of it in March 1874. Two years later in March 1876, further floods necessitated heightening of the embankment.

Despite this, the line was flooded again in February 1882.[84] Thereafter its height was raised progressively but the next flood always seemed to be worse than its predecessor so that whatever repairs were carried out, they were no enduring deterrent to the sea. As noted earlier, on 'Black Monday', 29 November 1897, Essex was hit by one of the highest tides ever experienced as a great storm – one of the most severe North Sea storms on record – hit the south east of England.[85] Embankments and sea walls failed in many parts; there was massive destruction in Margate, Kent, for instance and, in common with the Southwold Railway, the Wivenhoe to Brightlingsea line was almost completely submerged. A train which attempted to make its way through the flood became derailed and the passengers had to be extricated with the aid of boats.[86] Elsewhere in Essex, over 3 kilometres of line were flooded by the sea on the London Fenchurch Street to Southend-on-Sea route between Pitsea and Benfleet and the ballast and sleepers were washed away.[87] Another nearby coastal line in Suffolk, between Lowestoft and Oulton Broads, was washed away in the floods.[88] Although extreme, this event was just one of many to attack, inundate and rupture the Wivenhoe to Brightlingsea line.[89] Storms and floods in November 1901, November 1903 and December 1904 were especially problematic, causing extensive damage and closure of the line for repairs to the embankment. The Great Eastern Railway Company even considered closing the branch after the latter as it was proving so costly to maintain.[90] Still further floods and line closures took place in January 1928[91] and in April 1949.[92]

In places, the eroded remains of old wooden stakes driven into the mud near the base of the embankment are still present. These were just one of many early attempts at protection, designed to absorb the energy of waves breaking against the embankment. All of the previous flooding and storm events, from 1874 until 1949, were, however, to be totally eclipsed by the great floods of 1953 for which no-one was prepared. The devastation along the coasts of eastern and south-eastern England was immense; on the night of 31 January 1953 almost the entire length of the Brightlingsea branch was engulfed and washed out and, as a consequence, the line was closed immediately the following morning. Brightlingsea was an island again. Cattle and sheep were drowned, people were stranded in their homes and the flood waters extended up to 400 metres inland over the salt marshes. The scale of the devastation was such that it was feared the line may never re-open and that British Railways had been handed the perfect excuse for its closure. In places the tracks hung in mid-air where the sea had burst through, destroying the embankment and washing out the ballast. However, after some hesitation, the damage was made good over a lengthy period and the line eventually reopened to trains nearly 11 months later on 7 December 1953. It had valiantly survived the infamous East Coast floods; a decade later Dr Beeching was to kill it completely. The severe winter of 1962 to 1963 had nearly wiped out the town's oyster trade and passenger numbers were dwindling but it has been surmised that the main reason for closure was the escalating cost of maintaining the swing bridge over Alresford Creek.[93]

Today the old railway formation is no longer complete as the swing bridge at Alresford was demolished in 1967, shortly after the line's closure. The remainder of the disused track bed is now a coastal footpath with waves lapping at its base. The embankment has, following the 1953 floods and subsequently, been heavily strengthened by the emplacement of an apron of rock armour and it today acts as the sea wall (Figure 4.2), protecting the formerly reclaimed mudflats and

Figure 4.2 Looking north-west towards Wivenhoe from Brightlingsea along the track bed of the old railway line; today it is heavily strengthened by rock armour and acts as the sea wall. Sluices through the embankment permit drainage between the Colne Estuary and the marshes (photo: R. W. Duck).

salt marsh from natural tidal flooding. The marshes are now part of the Colne Estuary National Nature Reserve (NNR) but, thanks to the railway, their drainage and inundation cycles have been much modified despite sluices passing through the embankment. A local campaign to reopen the line, half a century after its closure, began in 2013.[94]

A tale of two estuaries

Although railway lines have isolated, for instance, Golant Pill, Bodmin Pill and Cockwood Harbour (as described in Chapter 1), and have straightened out the natural curvature of the coastline, these former embayments still have connections with the main water body by means of drainage culverts through their confining embankments. As such, though the patterns of water and sediment movement on the incoming and outgoing tides have been interfered with substantially, there are still linkages with the sea, albeit constricted. Elsewhere in Britain, railway construction has contributed to or caused even greater impacts at the coast, whereby estuaries have become completely blocked off from the sea and transformed into land.

It might be tempting to assume that the long embankment that carries the Ffestiniog Railway across the estuary of Traeth Mawr near Porthmadog in Gwynned was designed for that purpose. However, it was not. This structure was built in order to claim land for agriculture 25 years before the railway arrived to transport slate from inland quarries to the coast. William Madocks, landowner, entrepreneur and Member of Parliament, revived an earlier proposition that had been made in 1625 to Sir Hugh Myddelton to construct what is now referred to as the 'Cob'.[95] Myddelton, perhaps the most famous engineer of his day, had turned down a request from his cousin, Welsh baronet Sir John Wynn, to build an embankment, on the grounds of the size of his current workload coupled with his advancing years[96] Madocks, however, completed the project in 1811 and, cuckoo-like, the railway company took advantage of this embankment in 1825 to cross the estuary for its lowest mile into the port. A mighty and exposed structure, passing carriages have been derailed on it when caught by a sudden gust of wind.[97] It was breached by a gale once, in October 1927, when heavy seas broke through and caused widespread flooding. The damage was so extensive that the repairs took until the following February to be completed.[98] At this location, the railway cannot, however, be held responsible for the extensive coastal modification that has taken place.

Once connected with the railway network, the 'Welsh Riviera' resort of Tenby on the Pembrokeshire Coast was to become famous for its 'long stretches of golden sand and a glorious sea' along with its climate, 'admirably adapted for winter residence for the most delicate'. The Pembroke and Tenby Railway, which opened in 1863, provided the town with its first, albeit isolated, railway link to the west.[99] En route to Tenby from Penally, about 2 kilometres to the south west, the line traverses a straight embankment to landward of Tenby's South Beach. Here, sand and shingle pass landward into dunes. These are the famous Tenby Links, the birthplace of Welsh golf, on which the game has been played since about 1875.[100] On the landward side is the railway embankment which delimits the links as a narrow belt of land, about 350 metres in width at high tide. Still further to landward, on the north-western side of the embankment is an area of low-lying ground with hills to north and south. An area known as 'The Marshes' is today a large holiday park; nearby 'Marsh Farm' gives a further transparent clue as to the nature of the terrain along with 'Marsh Road', the section of the A 4139 between Penally and Tenby, that curves inland from the railway to cross the small River Ritec (Rhydeg) by means of Holloway Bridge.

Underlain by limestone of Carboniferous age, this is the lush valley of the Ritec, a stream that rises in the hills to the west of the village of St Florence, passes through the railway embankment by means of a culvert and drains through Tenby Links to the sea. The 1 inch to 1 mile First Series Ordnance Survey Map of this area, published in 1856, shows an estuary occupying this valley. Denoted as 'Holloway Water', it had a free connection with the open sea via a channel at least 150 metres in width at the entrance. In medieval times the estuary was navigable, at high tide, as far upstream as St Florence.[101] In the early nineteenth century the 'sheet of water, brought up by a full spring tide', was said to have extended inland 'a full mile'[102] In that day, the traveller between Pembroke and Tenby had to cross the estuary at low tide. Edward Donovan, author of *British Zoology*, passed this way in 1804. He recorded the fording of the Holloway Water in the second volume of his rambling epic, *Descriptive Excursions through South Wales and Monmouthshire in the Year 1804, and the Four Preceding Summers*:

There is at all times a narrow current through the marshes, along which the superabundant waters of an inland stream [the Ritec] are discharged into the sea, but at high water, when the flood is enlarged by the briny tribute of the Severn, the passage to Tenby by fording, as usual at other times, becomes impracticable, or at least dangerous to attempt.[103]

Donovan continued to note that: 'The foot passenger passes over the creek in perfect security when the water is at a moderate height, by stepping over a number of large stones that are placed across the bed of the stream for that purpose.'[104] This crossing place was along the line of the subsequently constructed main road over Holloway Bridge.[105]

So, was the construction of the embankment by the Pembroke and Tenby Railway Company responsible for the destruction of Holloway Water, by effectively closing it off from tidal inundation? Some half century before the arrival of the railway, between 1811 and 1820, long-serving Conservative MP, Sir John Owen of Orielton in Pembrokeshire, built a sea wall and embankment to keep out the sea.[106] With the aim of creating pasture by reclamation, the structure extended from Tenby to a place called Black Rock, which is approximately mid-way between Tenby and Penally and today forms an eminence in Tenby Golf Course. The outflow of the Ritec was permitted via a floodgate. Like others before and since, Owen's scheme suffered a setback in October 1826 when a storm broke though the embankment.[107] An 1830 navigation chart, surveyed by Lieutenant H. M. Denham R.N. depicts what is termed a 'causeway', partially constructed across the lower reaches of 'Holloway Marsh' and through which the river drains close to Tenby. Field boundaries drawn within the marsh are indicative of early land claim.[108] The actual sea wall was not, however, made good until 1840, by which time Sir John had relinquished ownership of the marsh. His original structure is said to be 'now under the railway'.[109] What is certain is that the railway – by enlarging, heightening and extending the embankment south westwards from Tenby to Black Rock to Penally – effectively finished off the job that Sir John Owen had begun of excluding the Ritec Estuary or Holloway Water from the outer reaches of the Bristol Channel. A tidal exclusion gate, upstream of the railway embankment, protects caravans and other properties from tidal intrusion. To the ecologist, the railway embankment has had a beneficial effect. It has turned part of the bed of the former upper and middle reaches of the estuary into 'Ritec Fen', a Site of Special Scientific Interest (SSSI) that extends for 2 kilometres on the north side of the river between Tenby and St Florence. Its mosaic of what are now wholly freshwater wetland habitats supports nationally important species of sedges, ferns and flies.[110] To the geomorphologist, who is concerned with evaluating coastal change through time, the closure of an estuary in this way was not only unusual but a highly significant occurrence, which has had a major influence on Tenby's South Beach for over a century.[111] By contrast, the historian of the Pembroke and Tenby Railway notes dryly that, 'Leaving Tenby the line runs behind sand dunes and a golf course to Penally.'[112] The demise of the Ritec Estuary due to the building of a railway line was not, however, unique in Britain.

The road sign beside the entrance gate to St Mary's Church, Brading, on the Isle of Wight denotes 'Quay Lane' (Figure 4.3), yet the spot is over 2 kilometres from the sea, at Bembridge Harbour, as the crow flies. The map of the island in *Bradshaw's Guide* of 1863, however, paints a very different picture. From Bembridge on the east coast of the island, 'The Traveller's & Tourist's Guide Map to the Isle of Wight' shows a large, wide estuary at the mouth of the River Yar, called Brading Harbour, extending inland to the village.[113] George Bradshaw described Brading as 'a decayed place . . . The harbour is like a shallow lagoon between Bembridge Point and St Helen's old chapel.'[114] This large expanse of water, also known as Brading Haven, has now gone,

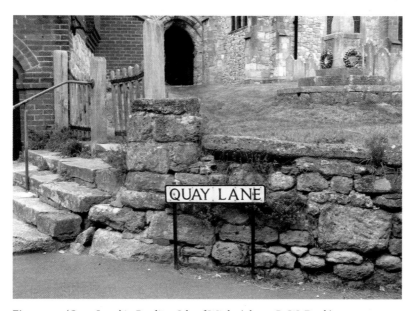

Figure 4.3 'Quay Lane' in Brading, Isle of Wight (photo: R. W. Duck).

Figure 4.4 View north eastwards across the Brading Marshes and the River Yar. The break in slope beyond the cattle marks part of the track bed of the former railway line to Bembridge (photo: R. W. Duck).

replaced today by the lush and verdant Royal Society for the Protection of Birds (RSPB) Brading Marshes Nature Reserve (Figure 4.4). Natural processes of sediment accumulation were not, however, the cause – the demise of this water body was due ultimately to a railway.

Until the late 1870s, at high tide cargo boats, albeit of shallow draft, could navigate the 4 kilometres from the open sea to the important island port of Brading and tie up alongside the quay at the eastern end of Quay Lane.[115] At low tide, vast areas of mudflats and salt marsh were exposed, incised by meandering creeks draining into the main artery of the River Yar. Whilst vessels had plied the route to Brading for centuries, an early attempt to reclaim Brading Haven prevented this happening for almost 10 years. However, it was to fail. In 1613, Sir Hugh Myddelton, goldsmith, entrepreneur and self-taught civil engineer, had transformed the sanitation of London by means of his New River that diverted clean water from the River Lea in Hertfordshire to Clerkenwell in the City.[116] Seven years later, in 1620, Myddelton's attentions became focused on the Isle of Wight when he built an embankment across the entrance to Brading Haven, extending from Bembridge to St Helens. This was not the first attempt to reclaim land, however; archaeological evidence points to a long history of earlier endeavours to win land from the sea in this area.[117] By means of Myddelton's embankment, some 800 acres of once inter-tidal land were no longer subject to tidal inundation. The structure was completed by 1622 but it was unable to withstand the relentless force of the sea and in 1630 it failed, resulting in all the reclaimed land being drowned and the abandonment of the project.[118] It might also have been the case that human intervention exacerbated the embankment's demise, as it was rumoured that local people had been removing earth materials from it to repair nearby roads.[119]

This experiment had cost in all £7,000, when the sea broke in ten years later, and Sir Hugh Myddelton's fields once more became harbour-bottom, and cockles and winkles once more grew where his meagre crops of oats and rape had struggled for existence.[120]

Moreover, sailing traffic to Brading resumed and cargo vessels continued to carry goods to and from the port for the next two-and-a-half centuries. The coming of a railway was to finally put this practice to an end – not in the way that the relative rapidity of railways killed off canal transport, but rather by the final closure of the estuary from the sea.

At the time of Bradshaw's 1863 guide, Brading was served by a 'main line' railway from Ryde in the north to Ventnor in the south,[121] most of which survives to this day as the 'Island Line'. The pier at Ryde began life in 1813 and is believed to be the oldest pleasure pier in England. It has been extended on several occasions to its present length of a little over 700 metres.[122] Immediately after the publication of the 1863 edition of *Bradshaw*, the structure was widened in 1864 to incorporate what was an originally horse-drawn tramway, which continued in service until 1969. In 1880, a third section was added to the pier[123] to carry the main railway right out into the English Channel to a northern terminus station at Ryde Pier Head, thereby providing a much more efficient connection with passenger ferries. Thus, the pier is essentially three piers in one; promenade, redundant tramway and railway (Figure 4.5). Proposals to construct a branch from Brading to Bembridge, then one of the most isolated places on the Isle of Wight, were also gathering momentum at this time. The enterprise of the Brading Harbour Improvement and Railway Company eventually came to fruition with

Figure 4.5 The view south towards Ryde on the Isle of Wight along what is essentially three piers in one. The closure and dismantling of the central tramway pier has left gaps between the promenade pier to the west and railway pier to the east. The 'Island Line' railway station at Ryde Pier Head sits above the waters of the English Channel some 700 metres from the shore (photo: R. W. Duck).

the opening of the line to the public in 1882.[124] The route skirted the northern perimeter of Brading Haven but, in order to swing eastwards from St Helens, affording access to Bembridge, it was necessary to build a large embankment that effectively dammed off the mouth of the Yar and once again reclaimed the extensive mudflats, 'which for years have only been haunted by the winkle picker.'[125] This was far from a straightforward task. Thousands of tons of chalk rubble and clay were brought from Bembridge Down and Portsdown Hill near Portsmouth and dumped across the mouth of the estuary,[126] thus excluding the sea from Brading Harbour once again. Sluices were installed to control the freshwater outflow of the River Yar. On 12 July 1879, the *Isle of Wight Observer* reported that:

> After many ominous shakes of the head from people residing in the locality of St Helen's, there can be no doubt that Mr Seymour [Mr F. Seymour was the contractor for the works] has successfully completed the task of keeping back the water from the extensive mud flats which for years have furnished such capital spots for the 'winkler' and the fowler.[127]

The newspaper's Brading correspondent should, however, have taken heed of those head-shaking sceptics who lived close to the embankment because, in a little over three months' time, it was to fail. That same broadsheet noted on 25 October 1879:

We hear with regret that the recent rough weather has seriously damaged the embankment which crosses the harbour from Bembridge to St Helens. The damage done is said to amount to several thousand pounds, and will not easily be repaired.[128]

The sea had, in fact, made a breach some 70 yards in width and Brading Haven had again become re-flooded: 'Vessels now come up and down to Brading Quay for chalk as if nothing had ever happened to check them.'[129] Furthermore, around the northern side of the harbour the newly-built railway was flooded and strewn with seaweed in several places. Mr Seymour entered into a new contract to rebuild the embankment at an estimated cost of £10,000. It was completed in 1880[130] and shortly afterwards, on 27 May 1882, the 4.5-kilometre branch was eventually opened to passenger traffic, whilst the 800 acres of reclaimed inter-tidal mud flats and salt marsh were reported to be 'rapidly becoming useful land'.[131] Indeed, in terms of man's nineteenth-century conquest of and contest for the coast, these lands had been 'redeemed'.[132] The branch line from Brading to Bembridge was seen as 'the consummation of a long series of intelligent effort and triumph over difficulty and discouragement'.[133] It had effectively reduced the surface area of Brading Harbour by some 80%. In 1898 the Embankment Road was built a little to the north of the railway, thereby advancing the defence seawards and reducing still further the size of the former estuary.[134] What is known today as Bembridge Harbour, at the mouth of the now heavily canalised River Yar, is all that remains of Brading Haven (Figure 4.6).

The engineering endeavours of a small railway company thus succeeded, though not without an expensive battle with the sea and some support from the nearby Embankment Road, where Sir Hugh Myddelton had failed. Never a particularly successful enterprise, dwindling passenger numbers meant that the branch line to Bembridge survived only until September 1953.[135] However, the embankment between St Helens and Bembridge endures to this day, along with much of the track bed around the former estuary. It is more than a little ironic that the RSPB, who today own and manage the Brading Marshes,

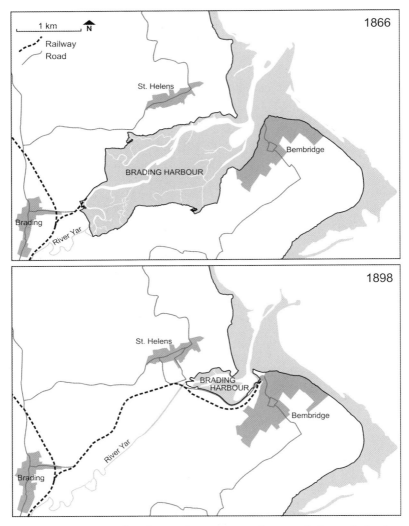

Figure 4.6 The extent of Brading Harbour in 1866 and in 1898, before and after the building of the railway. The line between Brading and Bembridge was opened in 1882; in 1898, as shown in the map, the Embankment Road was built immediately to the north of the railway (drawing by T. Dixon, based on maps © Crown Copyright and Landmark Information Group Ltd 2014. All rights reserved, 1866, 1898).

has recently proposed that the site should be inundated once more and restored to a more natural wetland environment.[136] The construction of our railways has had other, rather less obvious, impacts at the coast, as the following chapter explores.

Notes

1. 'The Wasting of the English Coast', *The Times*, 5 October 1886.
2. 'The Contest for the Coast' (1891), *Chambers's Journal of Popular Literature, Science and Art*, 8, 241–243.
3. 'London, Friday, May 5', *The Standard*, 5 May 1865.
4. Mendum, J. (2010), 'Henry Moubray Cadell: a geological and industrial innovator', *The Edinburgh Geologist*, No. 48, 5–14.
   Mendum, J. (2011), 'Henry Moubray Cadell and Lionel Wordsworth Hinxman: "elephants on Suilven"', *The Edinburgh Geologist*, No. 49, 7–12.
5. Mendum, J. (2010), 'Henry Moubray Cadell'.
   Cadell, H. M. (1929), 'Land reclamation in the Forth Valley. I Reclamation prior to 1840'.
   Cadell, H. M. (1929), 'Land reclamation in the Forth Valley. II Later reclamation and the work of the Forth Conservancy Board', *Scottish Geographical Magazine*, XLV, 81–100.
   Jehu, T. J. (1934), 'Obituary: Dr H. M. Cadell', *Nature*, 133, 822–823.
   'The power of the tides', letter to *The Scotsman*, 2 November 1920.
   'Flanders Moss Reclamation Scheme', letter to *The Scotsman*, 16 September 1926.
6. Cadell, H. M. (1913), *The Story of the Forth*, Glasgow: James Maclehose and Sons, 298 pp.
7. Cadell, H. M. (1888), 'The utilisation of waste lands', *Scottish Geographical Magazine*, 4, 366–379.
8. McLusky, D. S., Bryant, D. M. and Elliott, M. (1992), 'The impact of land-claim on macrobenthos, fish and shorebirds on the Forth Estuary, eastern Scotland', *Aquatic Conservation: Marine and Freshwater Ecosystems*, 2, 211–222.
9. McLusky, D. S., Bryant, D. M. and Elliott, M. (1992), 'The impact of land-claim'.
10. Davidson, N. C., Laffoley, D. d'A., Doody, J. P., Way, L. S., Gordon, J., Key, R., Drake, C. M., Pienkowski, M. W., Mitchell, R. and Duff, K. L. (1991), *Nature Conservation and Estuaries in Great Britain*, Peterborough: Nature Conservancy Council, 422 pp.
11. French, J. R., Benson, T. and Burningham, H. (2005), 'Morphodynamics and sediment flux in the Blyth Estuary, Suffolk, UK', in Fitzgerald, D. M. and Knight, J. (eds), *High Resolution Morphodynamics and Sedimentary Evolution of Estuaries*, 143–171, Dordrect: Springer.
12. French, J. R. and Burningham, H. (2003), 'Tidal marsh sedimentation versus sea-level rise: a southeast England estuarine perspective', *Proceedings of the International Conference on Coastal Sediments*, Clearwater Beach, Florida, USA, 18–23 May 2003, 1–14.
13. 'Coast erosion at Southwold', *The Ipswich Journal*, 8 February 1896.

Pye, K. and Blott, S. J. (2006), 'Coastal processes and morphological change in the Dunwich-Sizewell area, Suffolk, UK', *Journal of Coastal Research*, 22, 453–473.

Brooks, S. M. and Spencer, T. (2010), 'Temporal and spatial variations in recession rates and sediment release from soft rock cliffs, Suffolk coast, UK', *Geomorphology*, 124, 26–41.

14. 'Halesworth', *The Bury and Norwich Post, and Suffolk Herald*, 19 October 1875.

15. Jenkins, A. B. (1987), *Memories of the Southwold Railway*, Suffolk: L. & S. Rexton, 33 pp.

16. 'Opening of the Southwold Railway', *The Essex Standard, West Suffolk Gazette, and Eastern Counties' Advertiser*, 27 September 1879.

17. 'Opening of the Southwold Railway', *The Essex Standard, West Suffolk Gazette, and Eastern Counties' Advertiser*, 27 September 1879.

18. 'Opening of the Southwold Railway', *The Essex Standard, West Suffolk Gazette, and Eastern Counties' Advertiser*, 27 September 1879.

19. 'Great storm in Suffolk', *The Ipswich Journal*, 29 December 1894.
'The gale', *Jackson's Oxford Journal*, 29 December 1894.
'Coast erosion at Southwold', *The Ipswich Journal*, 8 February 1896.

20. Pennick, N. (1987), *Lost Lands and Sunken Cities*, London: Fortean Times, 176 pp.
Kington, J. (2010), *Climate and Weather*, Collins New Naturalist Library Book 115, London: HarperCollins, 486 pp.

21. 'The gale, Further loss of life and damage', *The Times*, 1 December 1897.

22. 'The gale, Serious loss of life at Margate', *The Times*, 3 December 1897.

23. Jenkins, A. B. (1987), *Memories of the Southwold Railway*.

24. 'Opening of the Southwold Railway', *The Essex Standard, West Suffolk Gazette, and Eastern Counties' Advertiser*, 27 September 1879.

25. Jenkins, A. B. (1987), *Memories of the Southwold Railway*.

26. Duck, R. W., McManus, J. and Diez, J. J. (1995), 'Comparative study of two largely infilled estuaries; the Eden Estuary (Scotland) and the Ria de Foz (Spain)', *Netherlands Journal of Aquatic Ecology*, 29, 203–210.

27. Duck, R. W. (2011), *This Shrinking Land: Climate Change and Britain's Coasts*, Dundee: Dundee University Press, 208 pp.

28. Maynard, C. (2014), Personal communication.

29. Robertson, C. J. A. (1974), 'The Cheap Railway Movement in Scotland: the St Andrews Railway Company', *Transport History*, 7, 1–40.

30. 'Opening of the St Andrews Railway', *Dundee Courier*, 7 July 1852.

31. *Fifeshire Journal* quoted in Hajducki, A., Jodeluk, M. and Simpson, A. (2008), *The St Andrews Railway*, Usk: The Oakwood Press, 288 pp.

32. 'Opening of the St Andrews Railway', *Dundee Courier*, 7 July 1852.

33. Rapley, J. (2007), *Thomas Bouch: The Builder of the Tay Bridge*, Stroud: Tempus Publishing Ltd, 192 pp.
McKean, C. (2006), *Battle for the North*, London: Granta Books, 390 pp.

34. Leighton, J. M. (1840), *History of the County of Fife from the Earliest Period to the Present Time, Volume III*, Glasgow: Joseph Swan, 264 pp.

35. Robertson, C. J. A. (1974), 'The Cheap Railway Movement in Scotland'.

36. Robertson, C. J. A. (1974), 'The Cheap Railway Movement in Scotland'.

37. Robertson, C. J. A. (1974), 'The Cheap Railway Movement in Scotland'.
38. Hajducki, A., Jodeluk, M. and Simpson, A. (2008), *The St Andrews Railway.*
39. Regulation of Railways Act 1868 c. 119 (Regnal. 31_and_32_Vict), available at http://www.legislation.gov.uk/ukpga/Vict/31-32/119/contents (last accessed 26 April 2014).
40. *Plan of St Andrews Railway from Milton to St Andrews*, November 1850, The National Records of Scotland, RHP 15549/1 and 15549/2.
41. Leighton, J. M. (1840), *History of the County of Fife.*
42. Leighton, J. M. (1840), *History of the County of Fife.*
43. Leighton, J. M. (1840), *History of the County of Fife.*
44. Leighton, J. M. (1840), *History of the County of Fife.*
45. Leighton, J. M. (1840), *History of the County of Fife.*
46. Fleming, D. H. (1897), *Handbook to St Andrews and Neighbourhood*, St Andrews: J. & G. Innes, 143 pp.
47. Haldane, R. and Buist, G. (1838), Parish of St Andrews, *Statistical Account of Scotland (1834–1845)*, 9, 449–498.
48. Hajducki, A., Jodeluk, M. and Simpson, A. (2008), *The St Andrews Railway.*
49. 'Terrific Gale', *Caledonian Mercury*, 1 November 1852.
50. *Fifeshire Journal*, 11 November 1852, quoted in Hajducki, A., Jodeluk, M. and Simpson, A. (2008), *The St Andrews Railway.*
51. Rich, F. H. (1864), 'North British, St Andrew's Railways', *Report of The Secretary of the Railway Department, Board of Trade*, 39–40.
52. Rich, F. H. (1864), 'North British, St Andrew's Railways'.
53. Rich, F. H. (1864), 'North British, St Andrew's Railways'.
54. 'Storm and Flood', *The Dundee Courier & Argus and Northern Warder*, 2 January 1877.
55. 'Storm and Flood', *The Dundee Courier & Argus and Northern Warder*, 2 January 1877.
    'The Storm', *The Star (St Peter Port)*, 2 January 1877.
    'Terrific Gale in the Channel', *The Standard*, 2 January 1877.
56. 'The Storm and Floods', *Northern Echo*, 2 January 1877.
    'Terrific Gale on the Coast', *Birmingham Daily Post*, 2 January 1877.
    'Terrific Gale in the Channel', *The Standard*, 2 January 1877.
    'Destructive Gales', *The Morning Post*, 2 January 1877.
57. 'Storm and Flood', *The Dundee Courier & Argus and Northern Warder*, 2 January 1877.
58. 'Storm and Flood', *The Dundee Courier & Argus and Northern Warder*, 2 January 1877.
59. *Fifeshire Journal*, 11 January 1877, quoted in Hajducki, A., Jodeluk, M. and Simpson, A. (2008), *The St Andrews Railway.*
60. Hajducki, A., Jodeluk, M. and Simpson, A. (2008), *The St Andrews Railway.*
61. John Milne, *Dictionary of Scottish Architects* (2013), available at http://www.scottisharchitects.org.uk/architect_full.php?id=200043 (last accessed 26 April 2014).
62. Reilly, C. W. (2000), *Mid-Victorian Poetry, 1860–1879: An Annotated Bibliography*, London: Mansell Publishing Ltd, 560 pp.
63. Wilson, J. H. (1910), *Nature Study Rambles Round St. Andrews*, St Andrews: University Press, 258 pp.

The West Sands Partnership, St Andrews (2011), *A Draft Management Plan for the West Sands of St Andrews, 2012–2025*, 16 pp.

64. Duffy, A. (2000), 'Bruce Embankment (St Andrews & St Leonards parish), watching brief', *Discovery and Excavation in Scotland (New Series)*, 1, 42.
65. John Milne, *Dictionary of Scottish Architects* (2013).
66. 'The Sea Encroaching at St Andrews', *The Scotsman*, 18 October 1898.
67. Fleming, D. H. (1897), *Handbook to St Andrews and Neighbourhood*.
68. Wilson, J. H. (1910), *Nature Study Rambles Round St. Andrews*.
69. Wilson, J. H. (1910), *Nature Study Rambles Round St. Andrews*.
70. Bruce Embankment Improvement Project Officially Opened in St Andrews (2005), available at http://www.fifedirect.org.uk/news/index.cfm?fuseaction=news.display&objectid=C35AB293-E7FE-C7EA-075828B39F008B6D (last accessed 26 April 2014).
71. McGlashan, D. J. (1997), *The Evolution, Environmental Effects of Coastal Protection and Suggested Strategies for the Future of the Eden Estuary, Fife, Scotland*, unpublished MSc Thesis, University of Strathclyde.
72. SUSCOD, available at http://www.suscod.eu/ (last accessed 26 April 2014).
73. StARLink – The St Andrews Rail Link Campaign, available at http://www.starlink-campaign.org.uk/page1/about.html (last accessed 26 April 2014).
74. Tata Steel: St Andrews Rail Link High Level Report (2012), available at http://www.scotland.gov.uk/Resource/0042/00422700.pdf (last accessed 26 April 2014).
75. British Railways Board (1963), *The Reshaping of British Railways, Part 1: Report*. London: HMSO, 148 pp.
76. Cinque Port Liberty Brightlingsea, available at http://www.cinqueportliberty.co.uk/index.htm (last accessed 26 April 2014).
77. Mills, A. D. (2003), *Oxford Dictionary of British Place Names*, Oxford: Oxford University Press.
78. Jones, R. (2012), *Beeching: The Inside Track*, Horncastle: Mortons Media Group Ltd, 131 pp.
79. 'Opening of the Wivenhoe & Brightlingsea Railway', *The Essex Standard and General Advertiser of the Eastern Counties*, 18 April 1866.
80. 'Wivenhoe & Brightlingsea Railway', *The Essex Standard and General Advertiser of the Eastern Counties*, 22 December 1865.
81. Brown, P. (1995), *The Wivenhoe and Brightlingsea Railway*, Romford: Ian Henry Publications Ltd, 158 pp.
82. Brown, P. (1995), *The Wivenhoe and Brightlingsea Railway*.
83. 'Wivenhoe & Brightlingsea Railway', *The Essex Standard and General Advertiser of the Eastern Counties*, 22 December 1865.
84. Brown, P. (1995), *The Wivenhoe and Brightlingsea Railway*.
85. Kington, J. (2010), *Climate and Weather*, Collins New Naturalist Library, Book 115, London: HarperCollins, 486 pp.
86. 'The gale. Further loss of life and damage', *The Times*, 1 December 1897.
87. Ransom, P. J. G. (2001), *Snow, Flood and Tempest: Railways and Natural Disasters*, Hersham, Surrey: Ian Allan Publishing, 176 pp.
88. 'The gale. Further loss of life and damage', *The Times*, 1 December 1897.
89. Fautley, M. and Garon, J. (2004), *The Essex Coastline – Then and Now*, Winterbourne Down: Potton Publishing, 272 pp.

90. Brown, P. (1995), *The Wivenhoe and Brightlingsea Railway.*
91. Ransom, P. J. G. (2001), *Snow, Flood and Tempest: Railways and Natural Disasters.*
92. Brown, P. (1995), *The Wivenhoe and Brightlingsea Railway.*
93. Brown, P. (1995), *The Wivenhoe and Brightlingsea Railway.*
    Ransom, P. J. G. (2001), *Snow, Flood and Tempest: Railways and Natural Disasters.*
    Jones, R. (2012), *Beeching: The Inside Track.*
    Holland, J. (2013), *Dr Beeching's Axe 50 Years On: Illustrated Memories of Britain's Lost Railways*, Newton Abbott: David and Charles, 192 pp.
94. *Gazette*, 25 July 2013, *Restore Brightlingsea to Wivenhoe railway line campaign launched*, available at http://www.gazette-news.co.uk/news/10572599.Resto re_Brightlingsea_to_Wivenhoe_railway_line_campaign_launched/?ref=rc (last accessed 26 April 2014).
95. Skempton, A. (2002), *Biographical Dictionary of Civil Engineers*, Volume I, *1500-1830*, London: Thomas Telford Publishing, 903 pp.
96. Davies, W. (1810), *General View of the Agriculture and Domestic Economy of North Wales; Containing the Counties of Anglesey, Caernarvon, Denbigh, Flint, Meirionydd, Montgomery*, Drawn Up and Published by Order of The Board of Agriculture and Internal Improvement, London, 510 pp.
97. 'Damage by the gale', *The Times*, 5 December 1914.
98. Ransom, P. J. G. (2001), *Snow, Flood and Tempest: Railways and Natural Disasters.*
99. Price, M. R. C. (1986), *The Pembroke and Tenby Railway*, Oxford: The Oakwood Press, 112 pp.
100. Tenby Golf Club, available at http://www.tenbygolf.co.uk/ (last accessed 26 April 2014).
101. Kelsall, D. (2003), *The Pembrokeshire Coastal Path: From Amroth to St Dogmaels.* Milnthorpe: Cicerone, 255 pp.
102. Harrison, W. (1966), 'Some aspects of Tenby's history', *The Pembrokeshire Historian: Journal of the Pembrokeshire Local History Society*, 2, 54–74.
103. Donovan, E. (1805), *Descriptive Excursions through South Wales and Monmouthshire in the Year 1804, and the Four Preceding Summers*, London: Volume II., Printed for the Author, 396 pp.
104. Donovan, E. (1805), *Descriptive Excursions through South Wales and Monmouthshire in the Year 1804.*
105. Harrison, W. (1966), 'Some aspects of Tenby's history'.
106. Harrison, W. (1966), 'Some aspects of Tenby's history'.
107. Harrison, W. (1966), 'Some aspects of Tenby's history'.
108. Bristol Channel, Tenby and Caldy Island (1830), Unnumbered Admiralty Chart of part of the Pembrokeshire coast, showing soundings, currents, coastal points, bays, lighthouses etc. Surveyed by H. M. Denham, Lieutenant R. N., Scale 9 inches to 4 nautical miles. Cut from a larger sheet, The National Archives, Kew, Reference: MFQ 1/1260/2.
109. Harrison, W. (1966), 'Some aspects of Tenby's history'.
110. Countryside Council for Wales, Ritec Fen, available at http://www.ccw.gov.uk/ landscape-wildlife/protecting-our-landscape/special-landscapes-sites/protected

-landscapes-and-sites/sssis/sssi-sites/ritec-fen.aspx?lang=en (last accessed 26 April 2014).

111. Thomas, T., Phillips, M. R. and Williams, A. T. (2013), 'A centurial record of beach rotation', *Journal of Coastal Research*, Special Issue 65, 594–599.

112. Price, M. R. C. (1986), *The Pembroke and Tenby Railway*.

113. *Bradshaw's Handbook for Tourists in Great Britain and Ireland* (1863), Section I – Bradshaw's Tours through the Counties of Kent, Sussex, Hants, Dorset, Devon, the Channel Islands and the Isle of Wight.

114. *Bradshaw's Handbook for Tourists in Great Britain and Ireland* (1863), Section I.

115. Brading: Official Town Guide, available at http://www.hiddenbritainse.org. uk/Downloads/projectmaterials/Brading_guidebook_low.pdf (last accessed 26 April 2014).

   Harding, P. A. (1988), *The Bembridge Branch Line*, Woking: Peter A. Harding, 32 pp.

   Kingshill, S. and Westwood, J. B. (2012), *Legends and Traditions from around the Shores of Britain and Ireland*, London: Random House, 528 pp.

116. Edwards, A. T. (1925), 'Great engineers – XI: Sir Hugh Myddelton', *The Structural Engineer*, 3, 371–376.

   Jenner, M. R. S. (2004), Myddelton [Middleton], Sir Hugh (1556x60?–1631), *Oxford Dictionary of National Biography*, Oxford: Oxford University Press, available at http://www.oxforddnb.com/view/article/19683 (last accessed 26 April 2014).

117. Brading Haven Old Sea Wall, Brading Isle of Wight: Level 1 Historic Building Record (2006), Prepared for Brading Town Council, WCA Heritage, Brightstone, Isle of Wight.

118. Edwards, A. T. (1925), Great engineers – XI: Sir Hugh Myddelton.

   Jenner, M. R. S. (2004), Myddelton [Middleton], Sir Hugh (1556x60?-1631).

   Harding, P. A. (1988), *The Bembridge Branch Line*.

119. Coastal Geomorphology, Technical Annex – Eastern Yar Strategy (2006), Atkins, 17 pp., and references therein.

120. 'Land Won from the Sea', *The Spectator*, 5 January 1895.

121. *Bradshaw's Handbook for Tourists in Great Britain and Ireland* (1863), Section I.

122. 'Improvement to Ryde Pier', *Isle of Wight Observer*, 9 July 1859.

   'An esplanade on the west side of Ryde Pier', *Isle of Wight Observer*, 8 October 1864.

   'Ryde Pier electric railway', *Isle of Wight Observer*, 12 November 1898.

   Bainbridge, C. (1986), *Pavilions on the Sea*, National Piers Society: History of Ryde Pier, available at http://www.piers.org.uk/pierpages/NPSryde.html (last accessed 26 April 2014).

123. Bainbridge, C. (1986), *Pavilions on the Sea*.

124. Harding, P. A. (1988), *The Bembridge Branch Line*.

125. 'Brading Haven Reclamation Scheme', *The Hampshire Advertiser*, 9 July 1879.

126. Harding, P. A. (1988), *The Bembridge Branch Line*.

127. 'Brading. The Reclamation of Brading Haven', *Isle of Wight Observer*, 12 July 1879.

128. 'Brading. The New Reclamation Works', *Isle of Wight Observer*, 25 October 1879.

129. 'Brading. The Reclamation Works', *Isle of Wight Observer*, 1 November 1879.

130. Maycock, R. J. and Silsbury, R. (1999), *The Isle of Wight Railway*, Usk: The Oakwood Press, 240 pp.
131. 'Bembridge', *Hampshire Telegraph and Sussex Chronicle etc.*, 22 July 1882.
132. 'The Contest for the Coast' (1891), *Chambers's Journal of Popular Literature, Science and Art*, 8, 241–243.
133. 'Bembridge', *Hampshire Telegraph and Sussex Chronicle etc.*, 22 July 1882.
134. Coastal Geomorphology, Technical Annex – Eastern Yar Strategy (2006).
135. Harding, P. A. (1988), *The Bembridge Branch Line*.
136. Booth, K. A. and Brayson, J. (2011), 'Geology, landscape and human interactions: examples from the Isle of Wight', *Proceedings of the Geologists' Association*, 122, 938–948.

5
Removing Shingle from the Beach is Prohibited

Builders have an affection for beach stones.
'Dungeness in danger', *The Standard*, 15 May 1883

Pebbles, Penzance and prostitutes

At the Penzance Petty Sessions on Monday, 21 July 1856, Joshua Burton, to whom the licence of the Seven Stars public house had recently been transferred, was fined £3 for allowing prostitutes 'to assemble' in that house. In the same court, Bartholomew Bottrell was fined a comparatively hefty £10 for removing shingle from the beach. The Bench, as on previous occasions, were determined to put a stop to this dangerous practice, 'a persistence in which will cause expenditure of vast sums in preventing encroachments of the sea.'[1] The plea that the shingle had been removed 'for the town' did not avail and Mr Millett, who prosecuted, expressed a desire to stop the practice.[2] Not deterred, Bottrell was prosecuted for the same crime again in August 1861 but, on that occasion, fined only 10 shillings as Lieutenant Peyton R. N., on behalf of the Admiralty, did not wish a heavy penalty to be imposed by the magistrates. The court heard that Bottrell had previously been fined for similar acts and it was thus intimated to him that any further offence would be more severely punished.[3] In the same month of 1861, five men were convicted of removing shingle from the beach between Penzance and Newlyn piers, contrary to Admiralty orders. They, however, were fined £10 between them.[4] Ten years later, in 1871, the Penzance court was still punishing this crime severely. William Friggins, Samuel Barnicoat and Samuel Hall of Penzance were each fined £10 plus costs for removing shingle from the beach in the parishes of Madron and Gulval, immediately to the east of the town, contrary to an order of the Board of Trade.[5] Later that same year, Mr Thorne, of the Hollies School, was fined £10 for procuring the removal of gravel from the beach for surfacing his garden paths.[6] No mention, incidentally, was made of any inappropriate 'assembly' at the Seven Stars on these subsequent occasions. But, why should what might appear as an innocuous crime merit a fine typically over three times greater in magnitude than that for keeping a house of ill repute? Was this merely a Cornish peculiarity? The Bench, in prosecuting Mr Thorne, defended the level of punishment in terms that removing shingle from the beach can endanger the safety of expensive sea walls and adjacent property. Moreover, this practice 'has been frequently dealt with severely here' in Penzance.[7]

The West Cornwall Railway arrived at its Penzance terminus in 1852, four years before Bartholomew Bottrell's first prosecution, the line into the town closely following the coastal edge for its final 2.5 kilometres. The railway station – built partially on land claimed from the beach – was opened in March but only a few months later, in December 1852, a portion of the sea wall protecting the line had been washed away.[8] This was the start of a protracted battle with the sea such that today a large apron of rock armour has been installed to protect the line from wave attack. Extraction of sediments can only have contributed to the lowering of the adjacent beach and thus the increased height of the waves breaking directly against the railway (Figure 5.1).

The magistrates of this part of Cornwall were, in fact, unusually enlightened. In 1609, an Act of Parliament of James I had been passed, '. . . for the taking, loading of sea sand for the bettering of the ground and for the increase of corn and tillage within the counties of Devon and Cornwall.'[11] Along the north coast of Cornwall in particular, this practice of fertilising the land with shell-rich sands led to the removal of vast quantities of material from local beaches. Sands from Bude, for instance, were a major source of supply; 4,000 horse-drawn loads are known to have been removed in a single day. In 1839 it was calculated that of the order of 5,600,000 cubic feet of sand were taken from the coast annually.[12] In many parts of south west England, however, this act was to become rather loosely interpreted by trades people, well beyond agricultural improvement, to include materials for roads, railways, ballast and the building industry. This was in contravention of an Act of George III that prohibited the removal of shingle or ballast where it would damage the shores, banks or rivers of the Kingdom including ports and harbours to a penalty of £10. Whilst the Crown owned the foreshore (see Chapter 1), the power of enforcing that Act in the nineteenth century was vested in the Board of Trade.[13] Penzance

Figure 5.1 The final approach to Penzance railway station at the end of the main line in Cornwall photographed in the storm of 5 February 2014. The waves are breaking against an extensive and substantial apron of rock armour installed to protect the railway, partially built on claimed land, from erosion (image copyright Roger Salter, Cornish Railways).

was not an exception; prosecutions for extracting beach materials took place the length and breadth of the county. In many parts, however, a blind eye was turned to the practice. In others, the Crown granted permission for the right to extract shingle in return for payment. For instance: '*Permission* has been granted in favour of Rev. D. King to raise gravel from the sea beach at Wemyss Bay, in the county of Renfrew, for an annual payment of one shilling'.[14] One can but speculate as to the state of upkeep of the pathways in the local church yard along with the garden of the nearby manse.

Sand and gravel are raw ingredients of the construction industry and their plentiful supply on many of Britain's beaches has proved an attraction for centuries. Thus, for time immemorial, carters have shifted sand and shingle from our beaches, whether or not it was legal or permission had been granted to do so. Such materials were enormously valuable for building purposes (Figure 5.2), making roadways, paths, foundations, embankments, ballast and as aggregates for the manufacture of concrete. Even Henry Cadell advocated the use of beach

Figure 5.2 Nineteenth-century building – the former Free School in Cromer, Norfolk – faced with small, wave-worn cobbles obtained from the beach and typical of so many buildings in the area (photo: R. W. Duck).

sand – obtained from around the shores of his native Firth of Forth and transported by rail – for the manufacture of concrete blocks for house building.[15] Shell sands, rich in calcium carbonate, were particularly prized as a source of lime for agriculture in many parts of Britain apart from Cornwall and Devon. Carters also removed seaweed from beaches to fertilise the land. This might seem a harmless practice but mats of dead and rotting seaweed provide a baffle against wave erosion and help to retain sand and gravel in place on the foreshore. Shingle was seen as a nuisance if it was thrown up by storm waves onto roads, promenades and other coastal properties. Furthermore, an attitude prevailed in coastal communities across the land that if shingle was allowed to remain untouched on a beach it would likely be encroached upon by a still greater enemy, the sea. It was therefore one's duty to remove it! However, there were also many people, from the length of Britain, who expressed their concerns about sediment extraction from beaches – from Peterhead, Aberdeenshire, and Broughty Ferry, Angus, in the north, to Sandown, Isle of Wight, in the south – citing the damage or destruction of local shoreline amenities, promenades and even houses as a consequence.[16]

The railways played their part in what was a burgeoning coastal extraction 'industry' during the mid to late nineteenth century. Beaches often provided a ready source of gravel and cobbles that could be used for building embankments and as track ballast. One-man operations on a local scale gave way to much more extensive extraction schemes. In many locations the rate at which sand and shingle was removed from beaches far exceeded the rate of replenishment. As a result, beach levels fell which, in turn, permitted waves of greater height than previously to break directly on the shore. With beaches depleted of their natural materials, erosion became enhanced and – especially along the frontages of expanding watering places – often developed into a serious problem. The reaction of the day was typically and not surprisingly to erect a sea wall to provide protection and the building material of choice was often concrete. For ease, convenience, and to minimise transport costs, it was not unusual in Britain for sea walls to be built using concrete mixed on site using locally-sourced sand and gravel aggregate. Indeed, these materials were acquired rather too locally since they were, more often than not, excavated from the immediately adjacent beaches, thereby lowering their levels even further. In this way, such a beach had already been lowered by sediment extraction – often very substantially – before the newly-constructed wall behind it had begun to induce any scour or undercutting at its toe, as sea walls are prone to do. Even stone blocks were, in some cases, sourced locally. Again, the railways were incriminated, since many lines were being laid, as at Dawlish, along the tops of mighty walls that were constructed right at the coastal edge.

'The Wasting of the English Coast'

On 7 October 1886 *The Times* published an anonymous and now long-forgotten article entitled 'The Wasting of the English Coast', apparently extracted from a report of the British Association.[9] It was subsequently reproduced in full two days later in the *Aberdeen Weekly Journal*.[10] It describes the rates of, causes of, and reactions to erosion along, in particular, the south coast of England from Cornwall to Kent. Although the 'sea is the common enemy' attitude prevails (see Chapter 4), this piece includes some very pertinent and perceptive observations concerning the impact of human activities at the coast that are just as relevant today as they were 130 years ago. As such, it deserves careful scrutiny. With regard to the widespread practice of removing beach materials *directly* – sand, shingle, rocks – for a multitude of purposes, the commentary points out that:

All along the coast we find the same tale of unchecked removal of material for agricultural, road-making and building purposes. Thus at Sidmouth, when the Board of Trade some years ago set up a claim to the foreshore, and forbade the removal of shingle, such a disturbance was made by the inhabitants that the attempt was given up; and now shingle, gravel for walks and roads, and sand for mortar are taken without let or hindrance. In some parts, as near Lyme, limestone is taken from the cliffs and ledges, and lime is copiously extracted from them. The very ledges of the shore are stripped off. From a tract east and west of the River Brit [which flows into the English Channel at West Bay below Bridport in Dorset], belonging respectively to General Pitt-Rivers and the Earl of Ilchester, and leased to private individuals, something like 10,000 tons may be carted away in six months. Between Littlehampton and Brighton it is said that most of the walls within two miles of the sea have been built with boulders from the beach; hence it is only natural that the shingle has been diminishing, as also has the broad tract of fine or mud sand. At Brighton the sands are clearly at a lower level than formerly.

The author was, however, at pains to point out that the extraction of sediments from beaches was not just an English south coast problem:

No one part of the country is much more guilty than another in respect of the removal of shingle, although some parts are more injured by it than others. The common answer to objections is that a single tide will sometimes bring in more shingle than a year's operations will take away; and that no real harm can be done by such abstraction. But it must be remembered that every existing portion of shingle is a product of destruction, and although its removal may not appear to make much impression, the same sea which destroyed cliffs to make that shingle will almost certainly destroy an equal portion to make up for that which is removed; and this process will go on faster in proportion to the increasing demands of mankind.

The *indirect* removal of beach sediments from the shore face of the source – typically via rapidly expanding port and harbour development works, which presented obstacles to the natural transport of sand and gravel, was also highlighted as a serious cause for concern, especially but by no means exclusively on the English Channel coast. Long before the 1915 landslides between Folkestone and Dover, the actions of the South-Eastern Railway Company were being questioned:

> The erection of stone harbour piers in many ports has been alleged as seriously aiding in the destruction of the coast. Folkestone is a typical case, and a connexion is believed to be made out between the elongation of the pier to the westward of the harbour and the denudation of East Wear Bay of shingle, causing an alarmingly rapid fall of the chalk under-cliff, to the imminent jeopardy of the South-Eastern Railway.

The latter company, whose network from London served the resorts and ports of the English Channel coast of Kent, was not only a victim of coastal erosion and landslide activity, induced at least in part by its own actions (see Chapter 3), but its extraction of shingle was also to become an important contributor to, or indeed cause of, coastal problems in that area.

The Craigie Beach Affair

A curious situation arose in Dundee that became known at the time as 'The Craigie Beach Affair'. In 1850 the proprietor of the Craigie estate on the port's eastern flank, Mr Kerr, had entered into an agreement with the Harbour Trustees permitting them to extract shingle from the beach. Subsequently, large amounts were removed for various purposes. Four years later, in 1854, the local Police Commissioners, believing that the shingle belonged to the Town Council, undertook to remove some for their purposes and, to that end, hired a barge from the Harbour Trustees to facilitate its transport. Similarly, the Dundee and Arbroath Railway Company, not content with marooning Dundee's nearby beaches (see Chapter 1), also made application to the Harbour Trustees for the use of a barge to aid the removal of shingle from Craigie Beach for railway ballast. The Harbour Trustees, however, subsequently claimed that, whilst they had granted the use of a barge to both parties and had taken rental payment for this, they denied having sanctioned the removal of any gravel, claiming neither did they know nor enquire for what purpose the barge was to be used. The Police Commissioners and the Railway Company had thus removed quantities of 'stuff' from the beach at Craigie without permission and, in consequence, an interdict was served not only upon them but also on the Harbour Trustees as they had facilitated their actions.[17] Following legal arguments, this was subsequently withdrawn unconditionally[18] but, during the period between 1850 and 1854, a report by the Harbour Engineer to the Harbour Trustees noted that a total 21,850 tons of gravel and boulders had been removed from Craigie Beach by all parties combined.[19]

A decade later the Police Commissioners were still being implicated in the extraction of beach materials, on this occasion in nearby Arbroath, Angus. On investigation of complaints made to Arbroath Town Council in 1864 about the removal of gravel and stones from the beach, it was found that 'the great defaulters were the Police Commissioners – (a laugh).'[20] It was found that no less than 1,000 bolls[21] of stones had been removed and piled up to be broken up for surfacing roads. One of the Councillors remarked that he thought it was a terrible thing that the Council should be allowing this to happen; the bulwarks of houses at the shore had already been partly washed away and if the removal of beach materials was to be permitted to continue, the houses themselves might be 'carried off.'[22]

Portobello tales

We are suffering now for years of carting sand from the beach for industrial purposes.

'Portobello Sands', *The Scotsman*, 13 September 1949

On John Ainslie's map of 1783, Portobello on the Firth of Forth near Edinburgh is denoted as 'The Village of Figget'.[23] Half a century later, the road that led from the village of Restalrig to the low-lying shore between Portobello and Leith was variously known as Figgat (or Figgate) Whins Road or Craigentinny Road. The early, horse drawn railway between South Leith and Portobello had made its mark in the area. It had, in particular, considerably altered the natural drainage of 'the wilderness of sand-blown downs, covered with furze [whins or gorse bushes], heath and wild grass'[24] that graded into the beach proper. However, the railway line did not, at least in its original mode

159

of propulsion, prevent access to the beach. The closure of the thoroughfare to the shore by the North British Railway Company was to take place in 1860, occasioned by the conversion of its line to steam power, as alluded to in Chapter 1. As a further element of complication, the following year the owner of Craigentinny House, through whose estate the road passed, imposed another intervening barrier closer to Restalrig in addition to that imposed by the railway company near to the shore. This consisted of a barricade of railway sleepers; moreover, in part, the roadway had been ploughed up to grow crops.[25] Three local people, Thomas Cowan and James Livingstone, both gardeners, along with Archibald Hare, contractor, declaring that this was not a private estate road but a right of way since time immemorial, took Miss Helen (or Ellen) Marsh of Craigentinny to court – and, furthermore, they won their case.[26]

The statements of the numerous witnesses called for the pursuers have common threads – that the road was widely used by many people of all walks of life and ages to gain access to the beach for various purposes; either travelling on horseback, driving horses and carts or as foot passengers. Moreover, the statements reveal that the removal of sand and gravel from the beach by the cartload, along with seaweed, was the custom and practice of the neighbourhood. Thomas Cowan, gardener, one of the pursuers, knew the Craigentinny road since he came to live in Restalrig, 25 years ago. He had frequently travelled on it, and was once challenged by the Craigentinny overseer when driving two carts of gravel to the Caledonian Dairy. Moreover, statement after statement made to the court by numerous witnesses underlined just how common this practice was:

My father, and my mother, after my father's death, were in the habit of sending down carts by this road to the sea beach for gravel.

I have seen large numbers of carts passing and repassing [sic] to and from the grass meadows. I have also observed carts coming from the beach laden with sand.

The road goes directly past Craigentinny House to the sea beach . . . Carts sometimes came up laden with sand, passing along by Restalrig and St Margaret's Well.

I have frequently seen carts on the road. I remember seeing carts filled with seaware [seaweed].

Most of the carts passing along the road were filled with sea-ware [seaweed], sand and gravel.

I went down the road for the purpose of carting sand and gravel from the sea-shore. The gravel was for the walls in Parsonsgreen.

I have seen carts coming along with sand, gravel and grass.

I have seen horses and carts with sand and sea-weed passing along the road.

I have driven several cartloads of wood by that road to Portobello, besides numerous loads of sand, gravel and sea-weed from the beach.

I have driven carts on the road about 100 times annually for ten years. I was for some time in the employment of Mr Miller. He knew that I carted the sand, gravel and sea-weed.[27]

Little wonder that the beach at Figgat Whins was becoming severely depleted in sediment. On instructing the jury to return a verdict for the pursuers, the part the railway had taken in blocking the right of way was also mentioned. The judge presumed that under the *Railway Clauses Act*, 'they should be able without any difficulty whatever to get the Railway Company to give them a level crossing over the line.'[28] Today this line is still open for freight and it is spanned by a bridge where the long-forgotten Figgat Whins Road to the beach once existed.

Immediately to the south east, Portobello was rapidly developing into a fashionable watering place to the extent that, in 1844, its then great stretch of white sands helped to earn it the accolade of the 'Scottish Brighton': 'In the evening the beach, crowded with gay promenaders, forms a gay and beautiful scene with the white sands stretching to a great expanse in unbroken smoothness; altogether this may be termed the Scottish Brighton.'[29] Indeed it grew swiftly to become the most fashionable watering place in Scotland, removed from the fumes of the City of Edinburgh[30] yet easily accessible from the capital and beyond by the North British Railway's network. The promenade itself along part of the frontage was initially constructed in 1860,[31] coinciding with the upgrading to steam of the South Leith branch railway. As an added attraction, a pleasure pier was completed in 1871,[32] furthering the popularity of the watering place. This structure is now largely forgotten, to the extent that it fails to get a mention in at least one major history of the British seaside pleasure pier.[33] What makes it particularly intriguing is that it was designed and built by none other than Thomas Bouch of, amongst others, St Andrews railway 'fame'[34] (see Chapter 3).

In the typical Bouch style, the pier was of light construction, of cast and wrought iron with timber decks. It was carried on screw piles driven deep into the beach sand and extended 1,250 feet from the shore so as to permit access by steamers. The deck was 22 feet wide, except at the pier head where it broadened out to accommodate the commodious Pier Saloon,[35] a place of entertainment and hospitality. As a consequence of a shortage of funds, Bouch took over the pier on its completion and ran it with his wife; for nearly 20 years it was a modest financial success but persistent storm damage made it expensive to maintain. On Thomas Bouch's death in 1880, the year after the Tay Bridge Disaster and by which time he had been knighted by Queen Victoria, Lady Bouch sold off the pier to the proprietor of a fleet of pleasure steamers on the Forth who, in turn, is said to have sold it to the North British Railway Company.[36] In 1917, during World War I, the structure was badly damaged in a severe storm. The exigencies of war dictated that its repair was a very low priority. Advertisements placed by the 'Galloway Saloon Steam Packet Company', the structure's final owners, inviting tenders for the 'Removal of Portobello Pier' appeared in *The Scotsman* in December 1917[37] and thus it was subsequently demolished and the scrap allegedly used to help the war effort.[38] It should, however, be remembered that Bouch's forgotten pier at Portobello survived for some four and a half decades; rather longer than his infamous and ill-fated Tay Railway Bridge.

In parallel with its growth as a fashionable resort, Portobello also had several prospering industries at that time and important amongst these was the manufacture of glassware. From 1834 onwards, until at least the turn of the twentieth century, this was largely devoted to the production of bottles.[39] The raw ingredient for glass manufacture is, of course, silica sand and it will probably be of little surprise to find that this was obtained from as close to the works as possible so as to minimise transport costs; that is from Portobello beach itself. The bottle factory was situated at the western end of the frontage where it received sand transported from the east by the predominant direction of longshore sediment migration. To some extent the industry flourished on the feed of sand towards it and on the extent to which the beach was to become depleted further to the east.[40] However, the very material that underpinned Portobello's attraction as a tourist resort became progressively denuded by the glass industry. In fact, the rate of extraction by far exceeded the natural rate of replenishment by marine processes so the beach level became progressively lowered. The 'Scottish Brighton' moniker became lost to the annals of time as the sea front and promenade became battered and eroded by waves of ever increasing height as the level of the beach adjacent to it fell.

Many local citizens were rightly concerned about the effect that industrial extraction was having on the beach. To attempt to dispel their unease, following a storm in November 1875, the Provost of Portobello, Thomas Wood, penned a letter to *The Scotsman* stating that the area of beach and adjacent promenade where most sand had recently been carted away had escaped injury from the storm. Thus, in his view, there was no link between erosion of the promenade and beach drawdown. Moreover, he accused his critics of exaggerating the poor condition of the beach:

> I deem it my duty to endeavour to prevent the prosperity of the town being affected by any misapprehension regarding the condition of our beach; the more especially as there exists a small coterie of dissatisfied persons who, although townsmen, appear to have a malign satisfaction in deprecating the amenity value of our beach . . .[41]

Wood further maintained in a subsequent letter that sand extraction was not the problem, rather the quality of the materials with which the promenade was made.[42] Pre-eminent geologist, Archibald Geikie, the Director of the Geological Survey of Scotland and simultaneously the first occupant of the Chair of Geology and Mineralogy in the University of Edinburgh, took the opportunity to visit the scene with his students. It was their first Saturday afternoon field excursion of the 1875 to 1876 academic session. Geikie, not surprisingly, expressed the opinion that the removal of sand could not fail to have been injurious to the beach. Furthermore, the sand, he said, 'acts as a barrier against the inroads of heavy waves, and, this being removed, the promenade is left dangerously exposed.'[43] Severe damage by storm waves continued to take place on a number of occasions between 1877 and 1896 and eventually it was necessary to rebuild the promenade and extend it over the entire one mile frontage of Portobello.[44]

'Proprietor', a correspondent to *The Scotsman*, made it obvious in no uncertain terms who he held responsible for the damage suffered in a storm on New Year's Day 1877; the Provost of Portobello, Mr Thomas Wood himself, who had permitted the excessive sand extraction. About a half mile length of the promenade was washed away in that event. 'As was to be expected, the late storm has made sad havoc of the Promenade Terrace here, exposed as it now is, by the removal of so much sand from the beach, to the action of a heavy body of water.'[45] He continued: 'It is not, however, my purpose to show how much Provost

Wood and his followers are to blame for the state the beach is in – those who are to suffer by it can decide that question themselves. My object is to draw their attention to the question of the future – namely, at whose expense the necessary repairs should be made.'[46] In a subsequent letter responding to the Provost's vehement denial[47] that he was in any way liable for what was no more than natural storm damage, 'Proprietor' added:

> It is true the properties on the beach now need protection against the waves, but this was not formerly the case, and I would ask Provost Wood who is to blame for the removal of the sand which formerly protected said properties? Certainly, if what has been repeatedly stated on this subject be true, he, Provost Wood, should be the last man to ask those proprietors to pay the expense rendered necessary by his conduct.[48]

However, sand continued to be extracted to fuel the glass industry. In 1911 the bad state of the beach was, yet again, remarked upon:

> For about thirty years at least the foundations of this magnificent beach have been gradually undermined by carting away the sand; and it is to be regretted that this practice has been allowed to go on for so long, for, instead as formerly a sandy shore facing the town, there is nothing but shingle and stones.[49]

The writer, furthermore, predicted that, if the vandalism of sand extraction continued, the foundations of the promenade would, without doubt, give way during a gale, especially at its western end.[50] This prediction was, indeed, to come true. With the demise of the glass bottle industry in the late 1920s, the building trade continued the custom and lines of horses and carts could be seen filled with sand and gravel leaving the beach at that time. By 1926 the beach level was as much as 2 metres below that of the promenade rebuilt 30 years previously.[51] The practice of sand extraction, however, is known to have continued until at least the mid-1930s by which time the state of the beach had been reduced to stones and muddy sand.[52] In July 1926, the promenade was extended westwards from the town: 'Amid cheers Lady Sleigh cut the ribbon and declared the West Promenade of Portobello open to foot passengers. She was presented with a pair of scissors and a 3d piece as a memento.'[53] However, this still left the original structure particularly susceptible to damage by storm waves. Moreover, sand was still being removed from the beach. Indeed only a few months later, in October 1926, when stormy weather hit the greater part of Scotland and caused havoc in the Firth of Forth, it was to succumb to the sea. About 150 yards of the old promenade was damaged close to Portobello town centre as it was battered by huge waves. Moreover, only a little over a year later, in December 1927, a similar length of the promenade was so badly undermined by the action of the waves that it had to be demolished.[54] This necessitated rebuilding in part and repairing and strengthening of the structure by means of a rock armour apron with sheet steel piles at the toe.[55] The impact of sand extraction continued to be felt after the cessation of this practice. In February 1937, a section of the promenade about 50 yards long was seriously undermined in a storm.[56] The white sands were no more and so the tourists went elsewhere. Following World War II, another set of costly repairs was required to protect the frontage of the town as 100 yards of the promenade collapsed in yet another storm.[57] Despite these repairs, the danger of flooding of properties along the promenade had by then become a real issue,[58] one that had not existed when there was a healthy beach in place. On Good Friday 1958, for instance, great waves crashed across the promenade at Portobello flooding several cafés and restaurants.[59] Something had to be done – the sand was not simply going to return to the beach at Portobello by natural processes; the situation at the Portobello seafront had become unsustainable.

In 1970, the feasibility of importing sand from elsewhere to 'nourish' Portobello beach was considered. In an attempt to restore the beach level to that prior to industrial-scale extraction of material, the operation of sand replenishment began in the summer of 1972. This was the first major beach nourishment scheme in Britain. A bucket dredger was moored offshore from Fisherrow, about 3 kilometres to the east of Portobello, in the so-called 'borrow area'. Some 160,000 cubic metres of sand were dredged from the area, ferried to Portobello by barges and then pumped ashore and onto the beach face. Six timber groynes, equally spaced along the length of the beach, were constructed to retain the sand on the beach and prevent its migration towards the west under the action of north-easterly gales that drive the predominant, westward longshore sediment transport in the area.[60] As at Bournemouth on the Dorset coast of the English Channel, where another large beach nourishment scheme also dates back to the 1970s, the beach at Portobello needs to have its sand replenished roughly every 10 years.[61] Overall, this scheme has been a great success story and it has restored the beach to its former level both physically and in terms of its recreational amenity value.

In the case of Portobello, the railways played a pivotal but indirect role in the removal of sand and gravel from the beach. They were the catalyst for the growth of the popularity of the watering place and for its associated industry. The latter was responsible for the destruction of the beach and the removal of its natural protective capacity against erosion. The promenade – a prerequisite of a fashionable watering place – was a product of the influence of the railway and its frequent destruction or damage was due to the excessive extraction of sand by the glass and building industries such that removal far outstripped natural supply. In other parts of Britain, however, the link between railways and the removal of sand and gravel from beaches was explicit.

For instance, when the railway from Maiden Newton to Bridport in Dorset was extended for 2 miles to West Bay in 1884, *The Bristol Mercury and Daily Post* heralded 'A new watering place for the west of England'.[62] This small coastal village, at the mouth of the River Brit, soon expanded into a fashionable resort. The railway line, however, closed to passengers in 1930 but remained open for the carriage of freight until 1962.[63] By far the bulk of the latter was shingle removed from the beach below the railway station and hauled up the slope, in the shadow of the iconic East Cliff, in carts drawn by teams of heavy horses (Figure 5.3). From the carts it was transferred at the goods yard into railway wagons and transported across the country. In World War II it was used, in particular, for military airfield construction.[64] Such was the volume of shingle extracted from the beach for the manufacture of concrete and other purposes that it sustained the operation of the West Bay to Bridport goods line for over three decades.[65] The transport of shingle was clearly more profitable than that of passengers.

The Spurn gravel trade

The Holderness coast of Yorkshire's East Riding is one of the most vulnerable to erosion in Britain. This region projects southwards from Flamborough Head to Spurn Head, a curious, elongate spit of sand and shingle that extends in the direction of longshore sediment transport as a slender arc into the mouth of the Humber Estuary. With an elevation of no more than 9 metres at most above sea level and in places no more than this distance in width, there is no more fragile or indeed

Figure 5.3 A team of heavy horses hauling shingle from the beach at West Bay in Dorset on 7 March 1957 en route to the nearby railway station. In the background is East Cliff (sandstone of Jurassic age). This beach was used for the title sequences of the BBC's *The Fall and Rise of Reginald Perrin* in the 1970s and made famous most recently as a principal exterior location for the 2013 ITV drama *Broadchurch* (Image JV-L3985 by Valentine & Sons, Dundee, reproduced courtesy of the University of St Andrews Library).

more striking strip of coastal land in Britain.[66] The structure has been breached or over-topped many times by stormy seas over the years and the communities at the end of the spit continue to be cut off from the mainland periodically. The most recent breach was in the storm surge of 5 December 2013. Natural processes of erosion have been exacerbated by the 'Spurn gravel trade'. Cobble and gravel extraction along this coast was once big business, peaking in the mid-nineteenth century with sloops being the main mode of transport. These important building materials were continually collected onto carts along the seaward side of the spit, taken over to the Humber Estuary shore and piled into heaps, awaiting loading onto the sloops as they came alongside at low tide.[67] During a north-westerly gale on 28 December 1849, a breach was made through the spit which in 1850 was about 320 yards in width and 12 feet deep at ordinary high water and was used as a short cut by small vessels. By the following year the gap had grown to 500 yards wide and

16 feet deep at high water.[68] There is abundant evidence to indicate that the excessive removal of gravel and especially cobbles was a major contributor to this breach, which occurred in an area where extraction was concentrated. In the 1840s extraction was quite simply being carried out at a faster rate than natural process could supply the raw materials. Furthermore, the wave energy absorbing capacity afforded by the protective veneer of gravel and cobbles had been taken away. Even today this fragile environment has not fully recovered from the effects of this trade; the cobbles and gravel had likely taken centuries to build up in commercial quantities and, still depleted in these materials, it presents a vulnerable target for wave attack.

An almost forgotten railway also played an important but far less well known part in the Spurn saga. This was a military line, detached from Britain's rail network. In around 1915, during the First World War, a light railway was built along the length of this slender piece of land extending the 6 kilometres from Kilnsea in the north to the tip of Spurn Head. This enterprise was instigated by the War Department who believed that low-lying and remote Spurn could be a potential site for an enemy invasion. The line was required primarily to facilitate the transport of materials required to build various military installations, including the foundations for coastal defence guns at the northern and southern ends, along with a 300-yard sea wall at Kilnsea.[69] The latter structure was deemed necessary to protect the village, barracks and Godwin Battery from the ravages of the sea, where rates of erosion typically resulted in losses of land of over 2 metres per year. The contractor for the railway line, for the Kilnsea sea wall and for the various military defence works along the spit, was C. J. Willis of Manchester and London. Willis was well known in the early twentieth century for dock and railway construction projects both at home and overseas. Immediately to the south of Kilnsea, Willis installed a siding projecting eastward from the 'main' line; it was about 75 metres in length and descended the short distance directly onto the open beach. This was to become aptly known as the 'Sand Siding'.[70] Its purpose was all too obvious; to facilitate the efficient removal and transportation by rail of locally-sourced beach materials for the manufacture of the concrete needed for the various installations along the spit, including the almost adjacent sea wall. Even though beach sediments were removed in large quantities, Willis was not contributing to the 'Spurn gravel trade' *per se* since he was, surely to his mind, simply acquiring the materials he needed to fulfil his own contract. The fact that, in so doing, the beach level was being lowered and the spit made even more vulnerable to wave attack was, at the time, not even a consideration given the exigencies of war. Willis was immune to prosecution; his actions had the blessing of the War Department as they were carried out in the defence of the realm. The line continued to operate until 1952.[71]

Gunpowder and chalk

On Thursday 19 September 1850, the citizens of Seaford in Sussex, along with a multitude of visitors who had arrived by special trains from Brighton, London and beyond, were to witness an unusual spectacle that had been seven weeks in preparation. *The Times* reported that, 'The proposed operation was of a peculiar description, and upon a vast scale, and excited a great deal of interest.'[72] The spectacular white cliffs of the Sussex coast extend from Brighton to Beachy Head near Eastbourne. They are broken only by the broad mouth of the River Ouse, to the west of Seaford, and that of the Cuckmere River to the east of Seaford Head. Reaching a greatest height of 165 metres at Beachy Head, these vertical walls of chalk are naturally receding at an average rate of typically between 0.5 and 1 metre per year.[73] The predominant direction of longshore transport of pebbles that comprise the beaches along this stretch of coast is from west to east under the prevailing winds that approach Sussex from the south west. The 'plan' at Seaford was to blast away a huge mass of the chalk cliff, which, in theory, would create a large rock groyne that would prevent pebbles being transported from the town's beach towards Beachy Head.[74] By this means, Seaford would retain materials on its beach and erosion would be prevented. The impact on beaches to the town's east was of no consequence.

People flocked in from all quarters, even though it was a weekday, and by all available means of conveyance. The operations were conducted by the Board of Ordnance, with local landowners contributing towards the cost. Fifty-five men of the Royal Sappers and Miners were involved in the carefully prepared explosion. A tunnel at a height of about 50 feet above high water mark was driven into the cliff for about 70 feet. This branched into two galleries giving the form of a 'T'. At the end of each gallery, a chamber was packed with 12,000 pounds of gunpowder; the galleries were subsequently stopped up and tamped with chalk rubble. Above these two charges, a further 600 pounds of gunpowder was sunk into each of three great vertical shafts and similarly tamped with chalk rubble.[75] This was to be no small explosion. With detonation to be achieved by means of voltaic batteries, it was arranged that the two large charges would be fired simultaneously with the three above a few seconds thereafter. The great explosion was scheduled to take place at 3.00 p.m. but the crowds had started to amass several hours previously. 'It was a little odd', noted *The Times*, 'to see cigars smoked over the very spot where a dozen tons of gunpowder lay waiting but a spark to make them spring up with an irresistible force.'[76] With the spectators removed to a safe distance, that force was unleashed about 12 minutes later than scheduled. The mass of chalk which fell was indeed much larger than had been expected and thought to have comprised around 300,000 tons:

> The cliff thrown down, and the place whence it came, were objects of much interest for some little time after the explosion; but the ground soon began to clear, and the principal inn became a scene of no little excitement. When the throng had obtained there what they could, the general movement was to the railway station; and, by means of extra trains sent out as fast as they could be got ready, a host of people were despatched to their homes well satisfied apparently with their day.[77]

The writer and poet Richard H. Horne, friend of Charles Dickens and sub-editor of the latter's journal *Household Words*, featured the scene in a short story, 'Gunpowder and Chalk':

> It was three o'clock – the hour of doom for the chalk in its contest with gunpowder. A bugle sounded, and movement of the sentries on top of the rock was discerned by the thousands of eyes looking up from the beach. Many, also, who were above, suddenly thought they could better their positions by moving further off. Below, on the beach, there was a hush of voices; not a murmur was heard. Everybody stood in his favourite attitude of expectation. All eyes were bent upon the lofty projecting cliff; and nearly every mouth was open, as if in momentary anticipation of being filled with an avalanche of chalk. Again a bugle sounded – and all was silence. Not a shingle moved.[78]

Natural process, however, immediately set to work on the chalk debris displaced from the cliff to the shore below. The large quantities of explosives used in the venture had blasted the chalk into very small fragments. These were very soon removed by waves and longshore currents and the enterprise, spectacular though it must have been for the many onlookers, was an equally spectacular failure.[79] Rather than retain shingle on the beach at Seaford, the blasting scheme, at least temporarily, accelerated the eastward supply of sediment.

The mouth of the Cuckmere River is located about 2 kilometres to the east of the site of the Seaford explosion. Even under natural conditions of sediment supply from wave erosion and cliff falls, it has a long history of becoming choked with gravel transported from the west by longshore currents.[80] Rising in the South Downs, this stream enters the English Channel at Cuckmere Haven, a natural gap in the chalk cliffs between Seaford and Beachy Head. In the 1930s, proposals were made to extract the shingle that was impeding the flow of the river and to sell this for building purposes.[81] To that end a light railway, with a gauge of 2 feet and unconnected to the network, was laid from the mouth of the Cuckmere River along its left bank to Exceat, some 2 kilometres to the north. The extraction industry commenced in 1933 and continued until 1964,[82] thereby helping to deplete the beach of its sediment for three decades and, in so doing, making it more vulnerable to wave erosion. The gravel was extracted from the beach by the East Sussex Transport and Trading Company using a dragline.[83] It was then transported by the railway in side tipping trucks to a wharf at the landward end of the line that facilitated loading into lorries for onward distribution. Even before the railway was built, in the 1920s and 1930s, flint boulders – known locally as 'blue boulders' – were collected from this and nearby beaches by gangs of men. These were used in the building trade and for various industrial processes.[84]

The desolate Crumbles

The East Sussex coast between Eastbourne and Pevensey comprises a huge bank of shingle known as the Crumbles. Today it has been encroached upon by retail parks, commercial and residential developments, and a large marina complex. Collectively, these have now either destroyed or obscure most of its natural form and have resulted in the loss of the shingle plant communities the area once supported. Furthermore, these developments have contributed to a reduction in the availability of sediment migrating eastwards by longshore transport along the beach face. The Crumbles was once a remote and desolate spot and the infamous scene of two grisly murders, which attracted widespread media attention at the time. In 1920 the body of a young

London typist, Irene Munro, was found buried in a shallow grave in the beach[85] and in 1924 the body of pregnant Emily Kaye was found in a lonely former coastguard's cottage.[86] A crime of a very different nature, however, was, perpetrated here by the London, Brighton and South Coast Railway. In 1862 an unfenced branch line, little more than a long siding, was laid through then open country to the Crumbles foreshore from Eastbourne, a distance of about 5 kilometres. Its principal *raison d'être* was to facilitate the extraction of shingle for use as rail ballast from this infertile terrain.[87] Indeed the branch was to become known simply as the 'Ballast Line', leading as it did to the 'Ballast Hole'. The railway company had negotiated the purchase of *not less than* 48,000 cubic yards of shingle from the local landowner, the Duke of Devonshire, at one (old) penny (c. 0.4 pence in today's post-1971 decimal currency) per cubic yard. Shingle extraction became a major operation for the company and this practice continued until 1932.[88] The scale of operation at the coastal edge was such that the London, Brighton and South Coast Railway made their line progressively more vulnerable to wave attack. The same gale that breached the railway embankment at Llanelli in January 1890 was just one event to have damaged the Ballast Line at the Crumbles.[89] Diplomatically preserving the anonymity of the company concerned, the authors of the foremost text on *Coast Erosion and Foreshore Protection* of its day noted wryly that: 'In one case a railway company made special rates to encourage the traffic in shingle from the foreshore adjoining their line, and some years after were compelled to go to great expense erecting sea-defences to protect their line from the encroachment of the sea.'[90]

Furthermore the extraction at the Crumbles not only starved beaches to the east of their natural supply of sediment but also contributed to the erosion of one of the London, Brighton and South Coast Railway's own main lines. The company's coastal route from Eastbourne to Hastings, which opened in 1846, hugged the coast for most of the way via Pevensey, Norman's Bay, Cooden Beach, Bexhill, Bulverhythe and St Leonards. In substantial stretches it was originally constructed on top of the uppermost shingle ridge, or berm, of the beach face. Thus it was particularly vulnerable to storms; even more so should there be any depletion in its supply of sediment from the west relative to that migrating eastward. The line succumbed to wave attack in a storm in October 1859 – along a particularly susceptible section between Bexhill and St Leonards – but it was soon repaired. In the meantime the passengers had to disembark, walk around the damaged section of line and board a train waiting for them at the other end.[91] Once shingle extraction had commenced at the Crumbles in 1862, the coast line to the east suffered further and more extensive damage during storms. In October 1870, for instance, the same section that was breached in 1859 – at Bulverhythe, between Bexhill and St Leonards – was again washed away.[92] Following repairs, this happened yet again in November 1875,[93] in the storm that also caused damage at Portobello. Only a matter of months later, in March 1876, the line was breached again near Bexhill, along a section 300 yards in length.[94] Along with widespread damage elsewhere in the country, including the promenade at Portobello, the Bexhill line was severed again on New Year's Day 1877.[95] The almost incessant wave damage to this part of the line was such that, by August 1881, it was deemed necessary to relocate it further inland.[96] Despite this, the line was still under threat in the 1920s[97] and today heavy rock armour defences protect it from the sea at strategic locations such as Cooden Beach and Bulverhythe (Figure 5.4). To what extent the shingle

Figure 5.4 The now heavily defended coastal railway line at Bulverhythe between St Leonards and Bexhill, East Sussex. The line was built on top of a gravel berm or storm ridge behind which buildings now occupy the lower, formerly marshy area – or swale – that was inundated when the berm was overtopped by storm waves (photo: R. W. Duck).

extraction by the London, Brighton and South Coast Railway exacerbated the damage to their coast line can never be ascertained but there can be no doubt that it was an important contributor.

Dungeness and the 'Second Railway King'

Some Shareholders almost laughed at him for taking their line to Dungeness. They bought a thousand acres at 5/- [£5.00] an acre, and laid the line. What had been the result? They had had their money back again in shingle, and over (hear, hear).

'The South-Eastern Railway', *The Standard*, 24 July 1891

George Hudson (see Chapter 1), the 'Railway King', died in 1871. His successor to this throne was arguably Sir Edward William Watkin. The 'Second Railway King' was a visionary entrepreneur and a Member of Parliament who had a diverse range of railway interests in Britain and overseas, including the chairmanship of several companies.[98] It was, however, his period as chairman of the South-Eastern Railway Company, from 1866 to 1894, that was to have an important and long-lasting influence on the Kent coast. The company's network served a populous suburban area to the south of London which, as its historian noted in 1895 following Watkin's resignation, 'touches most of the holiday resorts patronised by Londoners.'[99] Little wonder that many of the South-Eastern Railway Company's shareholders doubted Watkin's decision to buy land in one of the remotest parts of Kent and to open up a branch line, at considerable expense, to a spot at which, 'there were not ten men there within as many miles.[100] In 1879–80, Watkin negotiated the South-Eastern Railway Company's purchase of 1,000 acres of land (a little over 4 square kilometres), at a cost of £5.00 per acre, at Dungeness.[101] This triangular-shaped promontory on the Kent coast is largely a vast body of shingle – composed almost entirely of rounded flint pebbles, golden brown in colour. It is said to be the largest such mass in Europe and has been described as 'perhaps the only desert in England'.[102] The apex of this so-called cuspate foreland, which projects some 7 kilometres south eastwards into the English Channel, is the closest point on the English coast to France, just 34 kilometres south to Cap Gris Nez. In surface area, Dungeness contains nearly 22 square kilometres of exposed shingle that has, over the past 5,000 years, become built up as ridges.[103] Sir Edward Watkin had thus acquired around one-fifth of the area of exposed shingle in this remote part of the south coast of England. 'It is a weird place', wrote the correspondent 'Terra Firma' to *The Standard*:

> . . . compared with which Salisbury Plain is as the Garden of Eden. It is the Arabia Petrea [sic] of the South Coast, and the first business of anybody who gets there is to get back again. Yet, strange to say, Sir Edward Watkin has just opened a railway to Dungeness, being an extension from Lydd, on which he runs three trains up and down each day, except Sundays.[104]

In 1881, the South-Eastern Railway Company had acquired the Lydd Railway from Appledore to Lydd with the intention of extending the line to Dungeness. So what was the motive behind Watkin's outlay of £5,000 for shingle on top of the cost, approaching £7,000 per mile,[105] of building a 4.5-mile long (7 kilometres) line from Lydd to a station next to the remote lighthouse at Dungeness? Watkin's vision was that this single track route, which opened in 1881, could eventually become a part of a new route to the continent via the South-Eastern metals.[106] Furthermore, he envisaged Dungeness as an *inexhaustible* source of shingle that could be extracted on a vast scale and, importantly, the South-Eastern Railway Company would have the monopoly on its removal. He saw Dungeness as a wasteland, the very materials of which should be put to good use. Once extracted, these could provide ballast for the company's growing network of lines, aggregate for concrete, the fill for roads and railway embankments and a host of other building and construction needs. Removing the shingle directly from the beach via the new railway would ultimately create a huge pit which could, believed Watkin, become the most cheaply produced, deep water refuge dock on the south coast of England, indeed anywhere in the world – a new gateway by rail and short ferry crossing to France.[107]

Watkin's scheme aroused considerable concern in the press from the outset. An Editorial in *The Standard* remarked that, 'the proposal, mooted by Sir Edward Watkin, to transport large quantities of the Dungeness shingle by means of a new railway to the district served by the South-Eastern lines, is one to be viewed with some apprehension.'[108] The longshore transport of gravel is in an eastward direction along the Sussex coast towards Dungeness. If Watkin was to remove large quantities completely from the coastal system, the fear was that this could accelerate erosion further to the east as the natural supply of sediment would be reduced. Groynes, so numerous on the south coast of England, were seen as the root cause of many problems:

Hove stops a portion of the beach from going to Brighton, and Brighton is equally ungracious towards Eastbourne, which in turn intercepts the flow of the beach to St Leonards. So also St Leonards would place Hastings at a disadvantage, and finally Hastings is trying hard to catch all it can out of the stock that is left to it.[109]

Nonetheless, shingle was being sold from the shore at Hastings.[110] The fear was that the South-Eastern Railway Company would even further deplete that stock – and permanently so.

'Terra Firma' called for an urgent inquiry into the operations of the South-Eastern Railway at Dungeness and, of Sir Edward Watkin, declared that, 'something should be done so that man, by his mischievous meddling, shall not sacrifice to the greedy sea any portion of that area which properly belongs to our "tight little island."'[111] The counter view was that Watkin was being wise and foresighted in looking after the interests of the Company and its shareholders.[112] 'Terra Firma' retorted:

I contend it will be of considerable importance to know whether there is any risk that the natural transit of shingle along the foreshore past Dungeness will be interfered with. Dungeness itself may suffer no particular harm; but other places may receive damage if they find that the shingle which ought to come to their foreshore is intercepted at Dungeness.[113]

Needless to report, there was no inquiry. The material extracted would act as a further fuel to industrial progress. Initially, at least, passenger traffic was of little of no consequence to Watkin and the line was used solely to transport shingle.[114] Dungeness station, when it was eventually opened in 1883 to accommodate passenger traffic, was a primitive hut 'without booking office, signals, or any other of the usual appurtenances of ordinary railway management.'[115] The guard both issued and checked the tickets on board inbound and departing trains; an early portent of the mode of operation of many branch lines in Britain that survived into the post-Beeching era. The Dungeness branch was not, however, to be one of these; it was closed to passengers in 1937 but continued to be used to transport shingle and other goods until the early 1950s,[116] by which time the South-Eastern Railway Company was no more. In seven decades or so, the line conceived by Watkin shifted vast quantities of shingle from Dungeness. Not only would the South-Eastern transport shingle for collection to any of the goods stations on its network,[117] it supplied flint to the Staffordshire potteries and the aggregate for many major civil engineering projects in the Kent area including the new harbours at both Dover and Folkestone.[118] At the latter, the new pier was built 'exclusively of concrete blocks which are made on the spot from three kinds of material – shingle brought from Dungeness, sand from Higham and cement from the works on the Medway.'[119]

Watkin held an obsession with communication with the continent at the expense, according to many of the South-Eastern Railway Company's shareholders, of caring for the needs of domestic passengers. Moreover, he had another very specific requirement for the shingle from Dungeness. In 1872, a Channel Tunnel Company had been formed with the aim of building a tunnel under the English Channel to France.[120] Nine years later, in 1881, under the direction of Watkin, experimental borings commenced by the Submarine Continental Railway Company at the foot of Shakespeare Cliff between Folkestone and Dover. Here a shaft was sunk in the chalk to a depth of about 160 feet at the foot of which was a square chamber from which the tunnel projected. A party of dignitaries, including the Lord Mayor of London, visited the site on 18 February 1882. *The Illustrated London News* reported that the boring had by that time advanced to a length of 1,250 yards, a rate of 3 miles per year. Simultaneous borings from the French side at the same rate would complete the project in three-and-a-half years. The tunnel was to be lined with concrete, 2 feet in thickness, made of shingle brought, as one might have surmised, by rail from Dungeness and cement formed from the chalk excavated from the tunnel itself.[121] But, it was not to come to fruition. The Board of Trade sought an injunction forbidding Watkin to proceed further, on the grounds that he had infringed the Crown's foreshore rights by tunnelling. Watkin initially ignored the Board of Trade; various legal and parliamentary wrangles ensued but eventually the project was abandoned.[122] Thus, Dungeness's shingle reserves were spared the extent of depletion that might have been. Watkin's vision that Dungeness would become a major port was also to fade away as Dover and Folkestone grew in importance for cross-channel traffic. His successor as Chairman of the South-Eastern Railway Company, Sir George Russell MP, however, reported to shareholders at the half-yearly general meeting in July 1895 that Dungeness, 'with its magnificent harbour, should have a great future.'[123] By these words, those assembled at that meeting might have been forgiven for thinking that it had been built already. Watkin's venture was, however, never to become reality but the mining of shingle from Dungeness that he initiated has nevertheless left an indelible yet now largely forgotten mark on the south coast of England. It is impossible to ascertain just how much shingle has been removed but, one thing is for certain, the quantity was far greater than that from Spurn Head, Cuckmere Haven or the Crumbles and far greater than that replenished by natural processes. Furthermore, the sediment budget of the Kent coast of the English Channel has been severely and irreversibly depleted by Watkin's Dungeness scheme.

A 'king', a castle and a tramway

Scarcely a gale of any severity assails these shores without leaving its mark on Sandgate.

Frederick A. Talbot, 'The Constant Rivalry
of Sea and Shore', *The Windsor Magazine*, May 1899

The world wags still with an amiable slowness here.

C. G. Harper, *The Kentish Coast*, 1914

The transport of shingle along the Sussex coast is predominantly eastwards towards Dungeness; from Beachy Head to Eastbourne, to Pevensey Bay, to Bexhill, to St Leonards, to Hastings then across the county boundary into Kent. To the east of Dungeness the movement is similarly alongshore towards Folkestone, driven by the prevailing south westerly winds, with only local reversal of sediment transport on the east facing side of the Dungeness foreland. Yet another of Sir Edward Watkin's unsuccessful schemes once existed along the edge of the waterfront between Hythe and Sandgate, the latter now a western suburb of Folkestone. This involved the acquisition of the horse-drawn Hythe and Sandgate Tramway, the only example of a street tramway owned by a main line railway company – Watkin's South-Eastern.[124] The railway access to Folkestone Harbour (see Chapter 2) was very steeply graded and therefore slow. The route of the Hythe and Sandgate Tramway was one of several regarded as having the potential to be developed into a railway to improve the approach to this important port; hence Watkin's interest in it. It is also worthy of note that Watkin was the Member of Parliament for Hythe and Folkestone for 21 years, from 1874 until 1895.

The principal route was at the top of the sea wall directly along the promenade via Seabrook, close to the water's edge; it was about 5.5 kilometres in length and of standard railway gauge. Opened in 1891, having been built in several sections and extensions, the South-Eastern Railway Company acquired the tramway in 1893 and operated the cars with a stable of 25 horses.[125] Despite growing competition from motor buses, the tramway continued to operate until World War I when, in 1914, the horses were commandeered by the Government and the service was suspended. The service was resumed following the cessation of hostilities but finally succumbed at the end of the 1921 summer season to the competition of motor transport. The tracks were lifted in 1922.[126]

It should be remembered that Edward Watkin's shingle mining enterprise at Dungeness had begun in 1881, a decade before the tramway opened. It may be surmised that the supply of shingle in transit along the shore face towards Folkestone from Dungeness had, by that time, become starved. It may be purely coincidental but on 14 November 1894 – just three years after the tramway had become operational – it was badly damaged in a gale that caused much havoc along the south coast of England. *The Morning Post* reported that:

> At Sandgate the South-Eastern Railway sea wall was carried away for a length of 60 yards, and will cost hundreds of pounds to repair. The tram line is undermined, and it is feared that several weeks will elapse before trams will be able to run. The road was cut through within 3 ft. of the lifeboat house, which is now in a dangerous condition.[127]

The Standard estimated that the repairs would cost £700 and it was noted that, during the storm, the foam from the sea fell in the High Street like a snow storm.[128] The repairs were duly carried out but, not five years later, history was to repeat itself. Yet another huge storm – a south-westerly gale funnelling along the English Channel and coinciding with high tide – struck the coast on the night of 12 February 1899 and continued to rage for several days causing widespread damage. And yet again the Hythe and Sandgate Tramway was a victim but on this occasion the devastation was much greater. There were said to be three breaches in the sea wall, the longest of which was about 300 feet: 'the tram road for the whole of this distance had been undermined, and has entirely collapsed.'[129] The ripping out of the wall and roadway by the waves exposed broken water pipes and gas mains; the damaged sustained was extensive.[130] The tramway, however, was again duly repaired. There is no direct corroboratory evidence to associate the impacts of these two storm events on Sir Edward Watkin's tramway at Sandgate with his shingle extraction updrift at Dungeness. However, to infer such a link is almost inevitable.

In the same year that the Dungeness shingle operations commenced, 1881, Watkin's South-Eastern Railway Company acquired Sandgate Castle,[131] a coastal fortification built in the reign of Henry VIII. This was located close to the eastern terminus of the tramway and was an obstacle to the route being upgraded into a fully developed railway towards Folkestone. Though it was said to be worth £20,000, the aim was to demolish it: 'This step is rendered necessary by the extension of the Hythe and Sandgate Railway to Folkestone, the direction of which cannot conveniently be averted to clear the castle.'[132] However, this development, like several others of Watkin's, did not come to fruition. The sea, however, did its best to destroy the castle over the centuries; by means of numerous storms including the two mentioned earlier that damaged the tramway. Wave attack was no doubt exacerbated by the starvation of shingle and thus a fall in the level of the beach below the structure. In addition, the castle had been built originally, at least in part, from stone sourced locally from nearby inter-tidal reefs,[133] an ill-advised practice which must surely have also contributed to the increased likelihood of storm wave attack. It was repaired and partially rebuilt on numerous occasions over the centuries becoming particularly badly damaged in a gale in October 1923.[134] In 1927, the structure, 'much damaged by the sea in recent years' and by then surplus to railway interests was offered to the local authority by the Southern Railway Company – into which the South-Eastern had by that time merged – for £1,000.[135] It was ultimately disposed of the following year to a Mr A. Batchelor of Bleak House, Broadstairs.[136]

Epilogue

At the turn of the twentieth century the government of the day established a Royal Commission to assess, amongst other things, the nature, causes and magnitude of coastal erosion in the country. This august body published its findings in 1911 under the snappy title *Report of the Royal Commission Appointed to Inquire into and to Report on Certain Questions Affecting Coastal Erosion, the Reclamation of Tidal Lands, and Afforestation in the United Kingdom.*[137] Perhaps unsurprisingly, a key finding was that: 'The removal of materials from many parts of the shores of the Kingdom and the dredging of material from below low water mark, have resulted in much erosion on neighbouring parts of the coast . . .'[138] Nevertheless, despite our more enlightened attitudes and better appreciation of the consequences, the custom and practice of removing beach materials continued around our shores until well into the twentieth century. For instance, although sand extraction for agriculture had by then ceased in most parts of Cornwall, it was still continuing on a small scale at some sites in 2007.[139] Elsewhere the illicit removal of shingle still takes place from time to time, for instance at Ballantrae in Ayrshire in 2005[140] and from Rossall beach in Cleveleys on the Fylde coast of Lancashire as recently as 2012.[141] In some parts of the country, notices warning that, 'Removing shingle from the beach is prohibited' remain today (Figure 5.5), implying that the practice has not yet been fully eradicated. Railways, however, have long since ceased to be implicated in coastal shingle removal but the collective role they once played in denuding Britain's shores was substantial.

Figure 5.5 Sign beside the shingle beach at Inverbervie, Kincardineshire, behind which there is evidence of shoreline erosion and the installation of rock armour and concrete blocks as means of protection. The now closed Montrose to Bervie Railway (subsequently renamed Inverbervie) terminated close by this spot and for much of its route was laid along the raised beach, at the top of the low grassy slope in the photograph, that characterises this coastline. Locally acquired beach shingle was used as the rail ballast (photo: R. W. Duck).

Notes

1. 'Penzance Petty Sessions', *The Royal Cornwall Gazette, Falmouth Packet and General Advertiser*, 25 July 1856.
2. 'Penzance Petty Sessions', *The Royal Cornwall Gazette, Falmouth Packet and General Advertiser*, 25 July 1856.
3. 'Penzance Petty Sessions', *The Royal Cornwall Gazette, Falmouth Packet and General Advertiser*, 16 August 1861.
4. 'West Penwith Petty Sessions', *The Royal Cornwall Gazette, Falmouth Packet and General Advertiser*, 16 August 1861.
5. 'Removing shingle from the beach', *The Royal Cornwall Gazette, Falmouth Packet and General Advertiser*, 4 February 1871.
6. 'Penzance', *The Royal Cornwall Gazette, Falmouth Packet and General Advertiser*, 7 October 1871.
7. 'Penzance', *The Royal Cornwall Gazette, Falmouth Packet and General Advertiser*, 7 October 1871.
   'Local intelligence', *The Royal Cornwall Gazette, Falmouth Packet and General Advertiser*, 23 April 1852.
8. 'Local intelligence. West Cornwall Railway', *The Royal Cornwall Gazette, Falmouth Packet and General Advertiser*, 31 December 1852.
9. 'The wasting of the English coast', *The Times*, 5 October 1886.
10. 'The wasting of the English coast', *Aberdeen Weekly Journal*, 7 October 1886.
11. 'St Ives. Important Board of Trade prosecution', *The Royal Cornwall Gazette, Falmouth Packet, and General Advertiser*, 30 October 1869.
    Simmonds, A. and Frost, S. (1978), 'Conservation recommendations for a sand quarry at Gwithian Beach, St. Ives Bay, Cornwall', *Landscape Research*, 3, 17–18.
12. Ward, E. M. (1922), *English Coastal Evolution*, London: Methuen & Co., 262 pp.
13. 'St Ives. Important Board of Trade prosecution', *The Royal Cornwall Gazette, Falmouth Packet, and General Advertiser*, 30 October 1869.
    'Another prosecution for taking shingle from the coast', *The Hull Packet and East Riding Times*, 3 June 1870.
14. 'Bed or shores of the sea &c.', *Glasgow Herald*, 21 July 1858.
15. 'A concrete housing proposal', *The Scotsman*, 21 January 1919.
16. 'The removal of shingle from Peterhead South Beach', *Aberdeen Weekly Journal*, 23 March 1881.
    'Broughty Ferry Beach', *The Dundee Courier and Argus*, 14 February 1868.
    'Sandown', *Hampshire Telegraph and Sussex Chronicle*, 8 March 1873.
17. 'The Craigie Beach affair', *Dundee Courier*, 1 November 1854.
18. 'The Craigie Beach affair', *Dundee Courier*, 3 November 1854.
19. 'Harbour Trustees', *Dundee Courier*, 4 April 1855.
20. 'Arbroath Town Council', *The Dundee Courier and Argus*, 15 January 1864.
21. The 'boll' was the basic Scottish unit of dry capacity, which, though obsolete by the early nineteenth century, was still used informally into the twentieth century. It was used principally for grain, peas and beans – with regional variations – and not typically used as a measure of capacity for beach 'stones'. For grain, 1 boll was approximately equivalent to 6 bushels. The amount of stones extracted in this case was probably around 6,000 bushels; roughly equivalent to about 220 cubic metres.

22. 'Arbroath Town Council', *The Dundee Courier and Argus*, 15 January 1864.
23. Baird, W. (1898), *Annals of Duddingston and Portobello*, Edinburgh: Andrew Elliot, 509 pp.
24. Baird, W. (1898), *Annals of Duddingston and Portobello*.
25. 'Jury Court – First Division, The Craigentinny Right of Way Case', *The Scotsman*, 31 July 1861.
26. 'Jury Court – First Division. The Craigentinny Right of Way Case', *The Scotsman*, 31 July 1861.
27. 'Jury Court – First Division. The Craigentinny Right of Way Case', *The Scotsman*, 31 July 1861.
28. 'Jury Trials. Right of Way Case at Restalrig', *The Caledonian Mercury*, 31 July 1861.
29. 'Portobello', *The Caledonian Mercury*, 11 April 1844.
30. 'Portobello and its beach', *The Scotsman*, 12 September 1911.
31. Baird, W. (1898), *Annals of Duddingston and Portobello*.
32. 'Opening of the Portobello Pier', *The Scotsman*, 24 May 1871.
33. Bainbridge, C. (1986), *Pavilions on the Sea: A History of the Seaside Pleasure Pier*, London: Robert Hale, 221 pp.
34. Rapley, J. (2007), *Thomas Bouch: The Builder of the Tay Bridge*, Stroud: Tempus Publishing Ltd, 192 pp.
35. 'Opening of the Portobello Pier', *The Scotsman*, 24 May 1871. Portobello Heritage Trust, available at http://www.portobelloheritagetrust.co.uk/pier.html (last accessed 26 April 2014).
36. Rapley, J. (2007), *Thomas Bouch: The Builder of the Tay Bridge*.
37. 'Removal of Portobello Pier', *The Scotsman*, 22 December 1917. 'Portobello Pier', *The Scotsman*, 25 December 1917.
38. Rapley, J. (2007), *Thomas Bouch: The Builder of the Tay Bridge*.
39. Baird, W. (1898), *Annals of Duddingston and Portobello*.
40. Newman, D. E. (1974), 'Beach restored by artificial renourishment', *Proceedings of the 14th Coastal Engineering Conference, Copenhagen*, II, 1389–1398.
41. 'Damage to Portobello Esplanade', *The Scotsman*, 17 November 1875.
42. 'Portobello beach', *The Scotsman*, 22 November 1875.
43. 'The Portobello Beach', *The Scotsman*, 22 November 1875.
44. Newman, D. E. (1974), 'Beach restored by artificial renourishment'.
45. 'Portobello Promenade', *The Scotsman*, 16 January 1877.
46. 'Portobello Promenade', *The Scotsman*, 16 January 1877.
47. 'Portobello Promenade', *The Scotsman*, 18 January 1877.
48. 'Portobello Promenade', *The Scotsman*, 21 January 1877.
49. 'Portobello and its beach', *The Scotsman*, 12 September 1911.
50. 'Portobello and its beach', *The Scotsman*, 12 September 1911.
51. Newman, D. E. (1974), 'Beach restored by artificial renourishment'.
52. Newman, D. E. (1974), 'Beach restored by artificial renourishment'.
53. 'Portobello amenity. Notable improvement. Promenade extended', *The Scotsman*, 15 July 1926.
54. 'Portobello Promenade. Damaged by storm', *The Scotsman*, 28 December 1927.
55. Newman, D. E. (1974), 'Beach restored by artificial renourishment'.
56. 'Portobello Promenade. Damage by heavy seas', *The Scotsman*, 2 February 1937.

57. 'Storm damage at Portobello', *The Scotsman*, 20 January 1945.
58. Newman, D. E. (1974), 'Beach restored by artificial renourishment'.
59. 'Bleak start to the holiday', *The Times*, 5 April 1958.
60. Newman, D. E. (1974), 'Beach restored by artificial renourishment'.
61. Charlier, R. H., De Meyer, C. P. and Decroo, D. (1989), 'Beach protection and restoration Part II: A perspective of 'soft' methods', *International Journal of Environmental Studies*, 33, 167–191.
62. 'A new watering place for the west of England', *The Bristol Mercury and Daily Post*, 1 April 1884.
63. Atterbury, P. (2011), *Lost Railway Journeys*, Newton Abbot: David and Charles, 176 pp.
64. Jackson, B. L. and Tattershall, M. J. (1998), *The Bridport Railway*, Usk: The Oakwood Press, 224 pp.
65. Atterbury, P. (2003), *Just a Line from West Bay*, Bridport: Postcard Press, 84 pp.
66. Duck, R. W. (2011), *This Shrinking Land.*
67. Mathison, P. (2008), *The Spurn Gravel Trade*, Newport, East Yorks: Dead Good Publications, 22 pp.
68. de Boer, G. (1981), 'Spurn Point: Erosion and protection after 1849', in Neale, J. and Flenley, J. (eds), *The Quaternary in Britain*, 206–215, Oxford: Pergamon.
69. Hartley, K. E. and Frost, H. M. (1988), *The Spurn Head Railway: The History of a Unique Military Line*, Patrington: South Holderness Countryside Society, 51 pp.
70. Hartley, K. E. and Frost, H. M. (1988), *The Spurn Head Railway.*
71. Hartley, K. E. and Frost, H. M. (1988), *The Spurn Head Railway.*
72. 'The great explosion at Seaford', *The Times*, 20 September 1850.
73. May, V. J. (1971), 'The retreat of chalk cliffs', *The Geographical Journal*, 137, 203–206.
    McGlashan, D. J., Duck, R. W. and Reid, C. T. (2008), 'Unstable boundaries on a cliffed coast: geomorphology and British laws', *Journal of Coastal Research*, 24, 181–188.
74. Ward, E. M. (1922), *English Coastal Evolution*, London: Methuen & Co. Ltd, 262 pp.
75. 'The operations by the Sappers and Miners at Seaford', *The Times*, 14 September 1850.
76. 'The great explosion at Seaford', *The Times*, 20 September 1850.
77. The great explosion at Seaford. *The Times*, 20 September 1850.
78. Horne, R. H. (1851), 'Gunpowder and chalk', *Household Words*, II, 60–65.
79. Hutchinson, J. N. (2002), 'Chalk flows from the coastal cliffs of northwest Europe', in Evans, S. G. and DeGraff, J. V. (eds), 'Catastrophic Landslides: Effects, Occurrence and Mechanisms', *Geological Society of America Reviews in Engineering Geology*, 15, 257–302.
80. Ward, E. M. (1922), *English Coastal Evolution.*
81. Larkin, M. (2006), *In the Footsteps of Time: Geology and Landscape in the Cuckmere valley and Downs*, Uckfield: Ulmus Books, 83 pp.
82. Longstaff-Tyrrell, P. (2010), *Reflections from the Cuckmere Valley: 250 Years of Industry and Commerce*, Polegate: Gote House Publishing Company, 146 pp.

83. Mitchell, V. and Smith, K. (2001), *Sussex Narrow Gauge*, Midhurst: Middleton Press, 96 pp.
    Larkin, M. (2006), *In the Footsteps of Time*.
84. Larkin, M. (2006), *In the Footsteps of Time*.
85. 'London typist murdered', *The Times*, 23 August 1920.
86. 'Eastbourne crime', *The Times*, 5 May 1924.
    'Bungalow crime', *The Times*, 6 May 1924.
87. Mitchell, V. and Smith, K. (1986), *South Coast Railways – Eastbourne to Hastings*, Midhurst: Middleton Press, no page numbers.
88. Sussex Industrial Archaeology Society, 'Two branches and a siding', available at http://www.sussexias.co.uk/articles/branch_7.htm (last accessed 26 April 2014).
89. 'The violent gale', *The Times*, 24 January 1890.
    'Destructive gales', *The Standard*, 24 January 1890.
90. Owens, J. S. and Case, G. O. (1908), *Coast Erosion and Foreshore Protection*, London: The St Bride's Press Ltd, 144 pp.
91. 'Great storm', *The Standard*, 27 October 1859.
92. 'Epitome of news', *Cheshire Observer and Chester, Birkenhead, Crewe and North Wales Times*, 29 October 1870.
    'News of the week', *The Graphic*, 29 October 1870.
93. 'More disastrous floods', *Jackson's Oxford Journal*, 20 November 1875.
    'The weather and the floods', *The Ipswich Journal*, 20 November 1875.
94. 'Severe storms and floods', *The Morning Post*, 11 March 1876.
95. 'The Storm and Floods', *Northern Echo*, 2 January 1877.
    'Terrific Gale on the Coast', *Birmingham Daily Post*, 2 January 1877.
    'Terrific Gale in the Channel', *The Standard*, 2 January 1877.
    'Destructive Gales', *The Morning Post*, 2 January 1877.
96. 'Keeping out the sea', *The Standard*, 30 August 1881.
97. 'Coast erosion at St. Leonards', *The Times*, 19 November 1924.
98. Hodgkins, D. J. (2002), *The Second Railway King: The Life and Times of Sir Edward Watkin, 1819–1901*, Cardiff: Merton Priory Press, 713 pp.
    Sutton, C. W. (2004), Watkin, Sir Edward William, First Baronet (1819–1901), rev. Bagwell, P. S., *Oxford Dictionary of National Biography*, Oxford: Oxford University Press, available at http://www.oxforddnb.com/view/article/36762 (last accessed 26 April 2014).
99. Sekon, G. A. (1895), *The History of the South-Eastern Railway*, London: Railway Press Co. Ltd, 40 pp.
100. 'South-Eastern Railway Company', *The Standard*, 3 February 1879.
101. 'South-Eastern Railway Company', *The Standard*, 3 February 1879.
    'The South-Eastern Railway', *The Standard*, 24 July 1891.
    'Joint-Stock Companies. South-Eastern Railway', *The Morning Post*, 22 January 1892.
102. 'Dungeness-on-Sea: A conquest of the desert', *The Times*, 26 August 1935.
103. Scott, G. A. M. (1965), 'The shingle succession at Dungeness', *Journal of Ecology*, 53, 21–31.
    Hubbard, J. C. E. (1970), 'The shingle vegetation survey of Southern England: A general survey of Dungeness, Kent and Sussex', *Journal of Ecology*, 58, 713–722.

Findon, R. (1985), 'Human pressures', in Ferry, B. and Waters, S. (eds), *Dungeness: Ecology and Conservation*, Report of a Meeting Held at Botany Department, Royal Holloway and Bedford New College, 16 April 1985, 13–24.

May, V. J. (2003), 'Dungeness', in May, V. J. and Hansom, J. D. (eds), *Coastal Geomorphology of Great Britain*, Geological Conservation Review Series, No. 28, Peterborough: Joint Nature Conservation Committee, 754 pp.

Rigg, J. C. and Sage, R. C. (2006), 'Investigation into the impact of abstraction on epiphyte species at Denge in Dungeness Special Area of Conservation', *Quarterly Journal of Engineering Geology and Hydrogeology*, 39, 333–338.

104. 'Dungeness in Danger', *The Standard*, 15 May 1883.
105. 'The Lydd and Dungeness Railway', *The Times*, 4 July 1881.
106. Sekon, G. A. (1895), *The History of the South-Eastern Railway*.
107. 'London and the provinces', *The Standard*, 7 December 1881.
     'South-Eastern Railway', *The Pall Mall Gazette*, 24 July 1891.
     Sekon, G. A. (1895), *The History of the South-Eastern Railway*.
108. 'Keeping out the sea', *The Standard*, 30 August 1881.
109. 'Keeping out the sea', *The Standard*, 30 August 1881.
110. 'Coast erosion and reclamation: No. V. The South Coast', *The Engineer*, 15 June 1906.
111. 'Dungeness in Danger', *The Standard*, 15 May 1883.
112. 'Harbour works at Dover', *The Standard*, 17 May 1883.
113. 'The beach at Dungeness', *The Standard*, 18 May 1883.
114. 'London and the provinces', *The Standard*, 7 December 1881.
115. Harper, C. G. (1914), *The Kentish Coast*, London: Chapman and Hall, 373 pp.
116. Disused stations. Site record: Dungeness, available at http://www.disused-stations.org.uk/d/dungeness/ (last accessed 26 April 2014).
117. 'Dungeness in Danger', *The Standard*, 15 May 1883.
118. 'Harbour works at Dover', *The Standard*, 17 May 1883.
     Sekon, G. A. (1895), *The History of the South-Eastern Railway*.
     'The new harbour at Folkestone', *The Times*, 30 December 1899.
119. 'The new harbour at Folkestone', *The Times*, 30 December 1899.
120. Sutton, C. W. (2004), Watkin, Sir Edward William, First Baronet (1819–1901), rev. Bagwell, P. S. *Oxford Dictionary of National Biography*.
121. 'The Channel Tunnel', *The Illustrated London News*, 4 March 1882.
122. Sutton, C. W. (2004), Watkin, Sir Edward William, First Baronet (1819–1901), rev. Bagwell, P. S. *Oxford Dictionary of National Biography*.
123. 'The South-Eastern Railway', *The Standard*, 26 July 1895.
124. Lee, C. E. (1950), 'The Hythe and Sandgate Tramway', *The Railway Magazine*, October 1950, 698–700.
     Hart, B. (1987), *The Hythe & Sandgate Railway Incorporating the Hythe & Sandgate Tramway*, Didcot: Wild Swan Publications Ltd, 168 pp.
125. Lee, C. E. (1950), The Hythe and Sandgate Tramway.
126. Lee, C. E. (1950), The Hythe and Sandgate Tramway.
127. 'Great gale and floods', *The Morning Post*, 15 November 1894.
128. 'The gales and floods', *The Standard*, 15 November 1894.
     'Serious storms and floods', *The Huddersfield Daily Chronicle*, 15 November 1894.
129. 'The gales and floods', *The Standard*, 14 February 1899.

130. 'The disastrous storm', *Glasgow Herald*, 14 February 1899.
     'Sandgate sea-wall destroyed', *Daily News*, 14 February 1899.
     'Destructive floods', *The Morning Post*, 14 February 1899.
     'The great gale', *Lloyds Weekly Newspaper*, 19 February 1899.
131. English Heritage – Pastscape: Sandgate Castle, available at http://www.pastscape.
     org/hob.aspx?hob_id=465722&sort=2&type=&rational=a&class1=None&period
     =None&county=None&district=None&parish=None&place=&recordsperpage
     =10&source=text&rtype=monument&rnumber=465722 (last accessed 26 April
     2014).
132. Sandgate Castle, *The Morning Post*, 11 April 1882.
133. Ward, E. M. (1922), *English Coastal Evolution*, London: Methuen & Co. Ltd,
     262 pp.
134. 'Havoc at Folkestone and Hythe. Sandgate Castle wall breached', *The Times*, 13
     October 1923.
135. 'Sandgate Castle', *The Times*, 21 November 1927.
136. Untitled article, *The Times*, 6 March 1928.
137. Royal Commission on Coast Erosion and Afforestation (1911), *Report of the
     Royal Commission Appointed to Inquire into and to Report on Certain Questions
     Affecting Coast Erosion, the Reclamation of Tidal Lands, and Afforestation in the
     United Kingdom*, London: H. M. Stationery Office, 3 vols.
138. Royal Commission on Coast Erosion and Afforestation (1911).
139. Joint Defra/EA Flood and Coastal Erosion Risk Management R&D Programme
     (2007), *Sand Dune Processes and Management for Flood and Coastal Defence.
     Part 4: Techniques for Sand Dune Management*, R&D Technical Report,
     FD1392/TR, 49 pp.
140. 'The mystery of Ballantrae . . . who stole the shingle from the beach? Locals
     raise alarm as thieves target beauty spot', *Herald Scotland*, 20 April 2005,
     available at http://www.heraldscotland.com/sport/spl/aberdeen/the-mystery-
     of-ballantrae-who-stole-the-shingle-from-the-beach-locals-raise-alarm-as-thie
     ves-target-beauty-spot-1.56583 (last accessed 26 April 2014).
141. 'Campaigners in pebble theft plea', *The Blackpool Gazette*, 5 August 2012, avail-
     able at http://www.blackpoolgazette.co.uk/news/local/campaigners-in-pebble-
     theft-plea-1-4804042 (last accessed 26 April 2014).

6

A Little Exercise of Observation and Reflection

Strangest of all are the revolutions along the coast. Our seaport towns have been turned inside out. So infallible and unchanging are the attractions of the ocean that it is enough for any place to stand on the shore. That one recommendation is sufficient. Down comes the Excursion Train with its thousands – some with a month's range, others tethered to a six hours' limit, but all rushing with one impulse to the water's edge. Where are they to lodge? The old 'town' is perhaps half a mile inland, and turned as far away from the sea as possible, for the fishermen who built it were by no means desirous of always looking at the sea or having the salt spray blowing in at their windows. They got as far back as they could, and nestled in the cliffs or behind the hill for the sake of shelter and repose. But this does not suit visitors, whose eyes are always on the waves, and so a new town rises on the beach. Marine Terraces, Sea Villas, 'Prospect lodges', 'Bellevues', hotels, baths, libraries, and churches soon accumulate, till at length the old borough is completely hidden, and perhaps only to be reached by an omnibus.

'A little exercise of observation and reflection',
Editorial, *The Times*, 30 August 1860

Bradshaw knows not of this charming little place

There is no doubt that railways have had a profound, indelible and lasting influence on society in Britain since the mania of the Industrial Revolution. Mrs Sarah Gamp, the garrulous, umbrella-shaking, alcoholic midwife in Charles Dickens's *Martin Chuzzlewit*, went to the extent of blaming the excitement of mid-nineteenth-century train travel for inducing premature births:

I have heerd of one young man, a guard upon a railway, only three year opened . . . as is godfather at this present time to six-and-twenty

188

blessed little strangers, equally unexpected, and all on 'um named after the Ingeines as was the cause.[1]

Aside from Mrs Gamp's gin-stimulated fancifulness (but if the vibrations from passing trains can induce landslides, perhaps she had a point) railways have been an enormous force for change in a multiplicity of ways. Though not by any means exclusively, this is especially so at the coast. Aspiring watering places and resorts owed their burgeoning prosperity to the coming of the railway and without such means of communication there was little prospect of sustained growth and development. Very few watering places without a railway station prospered in their own right. A case in point in the mid-1880s was the would-be watering place of Southbourne-on-Sea in Dorset, not to be confused with the village of Southbourne in West Sussex. George Knowles had led an unsuccessful campaign to prevent the railway from reaching Scarborough 40 years earlier but the resort went on to flourish. However, as the London and South Western Railway had by-passed Southbourne to the north in 1885, a positive advantage was crafted of its lack of connection with the growing south coast network. It was originally developed by a local doctor as an aspiring rival resort to its larger and much more famous near-neighbour, Bournemouth. Today it has been all but swallowed up to become an eastern suburb of the latter. In 1885, a highly accomplished spin doctor made the bold assertion in *The Hampshire Advertiser* that:

Some months ago we mentioned the opening up of a new watering-place on the south coast by the construction of a railway to Swanage in the Isle of Purbeck. Southborne-on-Sea [sic] says The World deserves a word of notice now that it has just completed the construction of a fine esplanade, a third of a mile long, cut out of the cliffs, about nine feet above the beach level, and walled in so securely as to resist the most furious assaults of storms in the Channel. Bradshaw knows not of this charming little place, which is situated at the eastern extremity of Christchurch Bay, and commands fine views of the Channel and the Needles. It is about a mile from the picturesque old town of Christchurch, with its cathedral-like Priory Church, at the confluence of the Avon and the Stour, and is three and a half miles east of Bournemouth. Southborne has quite taken the lead of its older rival in this matter of an undercliff drive, and a pier will hereafter be added to the attractions of the place.

The absence of direct railway communication will serve to keep Southbourne select, whilst it is unlikely to repel the most desirable class of visitors. The undercliff drive offers admirable facilities for the erection of sheltered wintered houses, with a southern aspect, whilst the bathing accommodation is good, and the scenery varied and charming. The opening of the esplanade was celebrated on Thursday by a regatta, a dinner, and other festivities.[2]

Southbourne-on-Sea did indeed get its pleasure pier as predicted; construction began some three years later in 1888 at a cost of £4,000.[3] Sadly the added attraction was able to play its part in keeping the watering place 'select' for only a short period as the 300 feet long structure was very badly damaged by a severe gale in late December 1900. Over the next few days the structure was battered by continuing heavy seas, effectively putting to an end its useful life after just 12 years. Those responsible for the construction of the town's fine promenade and sea wall had also, like so many before them, clearly underestimated the stormy ravages of the English Channel as they too were almost completely destroyed by these events and left in a 'deplorable state'.[4] Eventually, owing to concerns about the safety of the public, the pier had to be demolished – ironically by the Bournemouth Corporation – in 1907.[5] Land and sea seldom live in perfect harmony. Where railways are or were once present, however, they have brought an added dimension and at many locations around our coasts the *physical* impact of their imposition on the edge of Britain, though numerically unquantifiable, has been profound.

'A new town rises on the beach': hindsight is a wonderful thing

Great, visionary engineers of the Victorian era such as Isambard Kingdom Brunel knew only of challenges, not problems. He was born of an age when the civil engineer was not only male but also king and nothing was impossible. In 1845 Sir John Rennie, towards the end of his inaugural address as the third president of the Institution of Civil Engineers, remarked grandiloquently:

When we look around us and see the vast strides which our profession is making on every side, and the deservedly high place it holds in public estimation, we cannot but feel justly proud; for without the slightest disparagement of the pursuits or studies of other professions, I may confidently ask, where can we find nobler or more elevated pursuits than our own; whether it be to interpose a barrier against the raging ocean, and provide an asylum for our fleets; or to form a railway, and by means of that wonderful machine – the locomotive engine – to bring nations together, annihilating as it were both space and time . . .[6]

Though he never held the Institution's presidency, Brunel's enterprises were the archetype of such 'noble pursuits', as the sea wall of the South Devon Railway bears witness. This coastal railway structure, along with others elsewhere in Britain, is a splendid symbol of Victorian self-confidence or, perhaps, over-confidence. It epitomises defiance, yet it was breached by the sea even before it had been completed. Furthermore, it has had to be repaired repeatedly and strengthened throughout its life, most recently – and, thanks to modern media coverage, most spectacularly – in the spring of 2014.[7] Brunel, his contemporaries and those engineers that followed him would not and could not be beaten by that old enemy, the sea. This line is now – thanks to Dr Beeching – the sole rail artery to Cornwall via Devon; it is a strategic passenger and freight transport route, yet it is perennially vulnerable to storm wave attack. And in that regard, as this book has shown, it is by no means alone. Had we been starting from scratch, with the benefit of hindsight and a much better understanding of the implications of doing so; many of our coastal railways would not have been built in the often very vulnerable locations that were chosen by our Victorian forebears. With its vulnerable and frequently breached low level, back beach stretches along with high level, cliff ledge sections, the Cumbrian coast line is a splendid case in point (Figures 6.1 and 6.2).[8] But because railways, even those long closed to traffic and with rails uplifted, have become integral to both the built and the natural environment of Britain for so long, we tend to overlook the physical impacts that they have initiated and continue to cause at the coast.

'You become even more aware of landscape', wrote Robert Macfarlane in his moving book *The Wild Places*, 'as a medley of effects, a mingling of geology, memory, movement, life.'[9] Railways are now – not only at the coast, but everywhere – fixtures; in the landscape and in our minds. They are part of that medley of effects, slicing through but now part of the tapestry of the landscape; ugly scratches once perhaps but now, with the passage of time, those wounds have healed. The time before railways came and the ruthlessness and rivalries of the companies involved in

Figure 6.1 The view to the north-east along the Cumbrian coast railway line towards Flimby on 13 February 2014. Rock armour has been newly installed to protect the line, which was severely damaged and breached by the sea during the storms of the previous month. In this section, the tracks were originally laid along the back of the beach and in the foreground is a remnant of an old sea wall originally built to provide protection. The area of grassland, on which the wind turbines are located, was formerly salt marsh. The floodwaters still evident in this area are the result of overtopping of the line repeatedly by storm waves (photo courtesy of N. Booth).

their construction are now completely lost to the memory. People can no longer remember a time, for instance, when railway embankments and sea walls did not impede or restrict access to the beach, or obscure the view. The beaches, inter-tidal flats and salt marshes that have been irrevocably lost to railway construction have similarly been long forgotten. Fetid, stagnant, stinking pools of sewage infested by venomous insects as a result of impounding by railway embankments that straightened the natural irregularities of the coast, have been cleaned up, landscaped or filled in. Disputes about land ownership and the imposition of barriers to former rights of access are now, to all intents and purposes, things of the past. Railway companies are no longer permitted to remove vast amounts of sand and gravel from beaches for ballast or the manufacture of concrete, thereby depriving them and their neighbouring beaches of their natural protection against wave erosion and

Figure 6.2 The view south along the Cumbrian coast railway line towards Nethertown station on 30 May 2013. One of the remotest in England, this is marked by the line of lamp posts immediately to the east of the house. Perched on a ledge on the cliff, this section of the line is particularly vulnerable to both wave attack and land sliding – new rock armour defence works are shown in the final stages of completion (photo courtesy of N. Booth).

perturbing natural patterns of longshore sediment transport. To some extent we have forgotten the enormous contribution that railways have made to the transformation of our coasts because it all happened so long ago. But the legacy of our coastal railways and, in particular, their physical impacts on the edge of Britain, lives on. Moreover, this is especially so as our climate changes – and it is indeed doing so.

As a result of railway construction, embayments, like Golant Pill, have been cut off from the sea and estuaries, such as Brading Haven, have been all but eradicated. Thus natural patterns and processes of sedimentation have been perturbed or interfered with. Embankments crossing inter-tidal lands and salt marshes have caused irrevocable damage, as in the Dovey Estuary. Furthermore, in many places the extent of land removed from natural tidal inundation has, through the passage of time, ceased to be obvious, as it has been drained and transformed into pasture. The natural linkage between cliff erosion and beach accumulation has been wholly or partially cut off by railways in many places. So too, the natural supply of sediment to the longshore transport system has been starved by the imposition of a massive protruding railway pier, as at Folkestone, or a railway along the foot of eroding or landslide-prone cliffs. Whilst the Dawlish to Teignmouth coastal edge is a prime example of the latter, there are many others.

For instance, the Pembrokeshire Coast Path around the shore of Saundersfoot Bay passes through a series of three tunnels near Coppet Hall between Wiseman's Bridge and Saundersfoot. These are the most obvious remains of a railway, the Wiseman's Bridge Branch of the Saundersfoot Railway, which was constructed in the early 1840s to carry coal.[10] This coastal section was built on top of a sea wall – much damaged by the waves over the years and repaired on many occasions – that today forms the popular coastal path. The tunnels took the single track line, which ceased to operate shortly after the outbreak of World War II,[11] through a series of rocky promontories. Elsewhere it was located on top of the wall built along the back of the beach, at the foot of cliffs and, at high tide, right at the water's edge. The rocks of this coast comprise folded, mixed sedimentary sequences of Carboniferous age, the harder and thicker units – typically sandstones – forming the headlands. Though the rocks are older and the railway has long closed, it is like Dawlish in miniature, but far less well known. And, just as at Dawlish, this sea wall has periodically suffered severe damage during winter storms.[12]

About 2 kilometres to the north east of Wiseman's Bridge, from which location this line formerly curved inland to Kilgetty, is the small village of Amroth. Waves drive sediment transport along the shore dominantly in that direction, around the sweeping curve of the bay towards Amroth and beyond to Pendine Sands. Amroth sits right at the coastal edge and is today particularly susceptible to erosion. This was first manifest along the Saundersfoot coast in October 1896 when a storm, which also inflicted great damage to the railways at Tywyn, Llanelli and to the sea wall under construction to carry the metals at Rhyl, caused extensive flooding.[13] It was a headline in *The Times*, however, that left no doubt as to the settlement's acute vulnerability: 'Welsh coast erosion: Seaside village to be abandoned'. What might, however, come as a surprise is to find that this appeared in print in April 1936, long before such actions are commonly thought of as being contemplated: 'Amroth, a picturesque village consisting of about 20 houses and standing on the South

Pembrokeshire coast between Saundersfoot and Pendine, is to be abandoned to the encroaching sea.'[14] The article noted that it had suffered
severely from coastal erosion for many years but a scheme of defence
works would 'entail an expenditure of about £30,000, while a further
£4,000 to £5,000 would be spent on a contract which had been recently
considered.'[15] Though the term was not in use at the time, this was an
early instance of a proposal for 'managed retreat' at the coast on the
basis that the cost involved in defending the village was not economic.
An expenditure of £30,000 in 1936 equates with about £1.8 million in
2014.[16] Amroth was not, however, abandoned in this way but a little
over two years later, at the end of July 1938, it came under renewed
threat from the sea, as waves washed away sections of a timber wall in
the process of being built to protect the village.[17] In September 1957,
the recently repaired defences failed yet again; Amroth was effectively
cut into two by an exceptionally high tide and the sea swept boulders
across the coast road as waves broke over the tops of houses.[18] A repeat
of these events occurred in another severe storm in January 1974, said
to have coincided with one of the highest tides for 300 years.[19] The principal reason for Amroth's vulnerability to erosion, today ameliorated
by groynes[20] and the artificial recharge of beach materials, is due, quite
simply, to the Saundersfoot Railway. Prior to its construction, a plentiful
supply of sediment from the cliffs ensured a substantial beach in front
of the village that naturally protected it from wave attack. However, in
much the same way that Dawlish Warren has been starved of its supply
of sediment from the local cliffs by the Dawlish sea wall, so too the wall
between Saundersfoot and Wiseman's Bridge that carried this long forgotten colliery line has had a similar effect.

The Fairbourne Steam Railway began life as a 2 feet gauge horse-
drawn tramway in 1895. It was converted to steam in 1916 and a gauge
of 15 inches. Rebuilt in 1984 with a gauge of 12¼ inches, this tourist
attraction carries passengers along the 3.2 kilometre long line from
Fairbourne village to Barmouth Ferry Station.[21] From the latter, a ferry
connection is available across the mouth of the Mawddach Estuary
to Barmouth. The northern terminus is located right at the tip of a
narrow sand spit that projects across the mouth of the estuary and
curves inland, sickle-like, towards Barmouth Bridge (see Figure 2.6). It
is thus located on a vulnerable coast, exposed to westerly gales. So too
is Fairbourne itself. Located right at the coastal edge on a low platform
directly beneath the cliff-top village of Friog, this area is extremely
susceptible to coastal flooding (see Chapter 2). Fairbourne's main line

railway station is on the Cambrian Coast line to the north of the site of the Friog accidents. Following the storms of January and February 2014, which caused so much damage in west Wales beyond the breaching of the Cambrian Coast railway, Fairbourne has gained notoriety. It has been named as one of some fifty communities in Wales alone that are likely to prove too expensive to defend from the sea in the future. As such, it is expected to be subject to managed retreat in 2025 when the local authority will no longer continue to maintain sea defences. It is anticipated that more than 400 homes will have to be abandoned in the village by 2055.[22] Not surprisingly, this has provoked hostile reaction from the local residents who are considering taking legal action.[23] Ironically, but of no consolation to the people of Fairbourne, it is the railway that is – at least partially – responsible for this situation. The 'bastard shaly rock' that should naturally weather from the cliffs to the south and its debris tumble to the beach below is now prevented from doing so by the railway line and its associated defences. So too, wave erosion at the base of the cliff is now restricted. Thus, the northward longshore sediment transport feeding the spit on which the Fairbourne Steam Railway is located is now starved of at least part of its natural supply. The sea-front houses of Fairbourne have also become ever more vulnerable to wave attack as the beach is no longer receiving its natural quantity of sand and shingle. Similarly, one coastal railway might potentially be contributing to the eventual demise of another.

Changing climate and rising sea levels

It is beyond the scope of this book to give a fulsome account of the continually increasing reservoir of evidence that supports global climate change and in particular global warming. In the context of the British coast, the reader is referred to *This Shrinking Land* and, in the wider sphere, to the most recent reports (2013 and 2014) of the Intergovernmental Panel on Climate Change (IPCC)[24] and the Geological Society.[25] As a consequence of global warming, due to increasing concentrations of carbon dioxide in the atmosphere, which leads to melting of ice sheets, ice caps and glaciers, along with the thermal expansion of the oceans, the IPCC has estimated that global sea level rose by an average of 1.5 to 1.9 millimetres per year between 1901 and 2010, a trend that has been accelerating since the coal-fuelled Industrial Revolution began in the late eighteenth century. Between 1993 and 2010, the rate of global sea level rise is estimated to have increased to between

2.8 and 3.6 millimetres per year.[26] This does not, however, mean that the whole of Britain's coasts are currently experiencing these amounts of sea level rise – there are considerable variations along the length of the country. A subtle, 'land-sea balancing act' is at play and what is critical is the 'relative' change in sea level at a given locality. The term 'relative sea level' refers to the balance between global sea level rise (or fall) and the vertical movement of the land, either upward or downward, at any particular locality. So, why might the land level change?

In Britain this is principally due to the fact that the country was subjected to 'ice-age' conditions during the Pleistocene epoch, between around 2,000,000 to 10,000 years ago. The thickness of ice was far from uniform, varying from its greatest in the Highlands of Scotland, where the precipitation is highest, to complete absence in the southern counties of England. Ice masses act as a dead weight, acting vertically downwards upon the brittle rocks that comprise the Earth's crust, causing it to be compressed under their great weight, pushing it downwards into the plastic part of the underlying upper mantle, which is capable of flow. This phenomenon is called isostatic depression. As this sinking takes place, mantle material becomes deformed and squeezed sideways from below an ice mass in much the same way as putty. This sideways movement of plastic mantle material away from below the ice pushes the crust upwards in areas not covered by ice, for instance, the south of England, making the land rise in what is known as a forebulge.

When the climatic conditions eventually improved and the ice masses began to melt at a greater rate than new ice was being formed, so the sea level around Britain began to rise. The sea did not, however, simply return to its pre-Pleistocene level. Because of isostatic depression of the crust and with large regional variations, vast areas of former coastal lowlands became flooded in what was known as the Flandrian Transgression, which reached its greatest extent about 5,000 years ago. However, the compression of the crust into the mantle beneath by the great weight of ice resting upon it had created an unstable situation and once the weight of ice was removed due to melting, the crust began to return to its former level, a process that is called isostatic recovery. It did so by rebounding upwards, moving slowly but discontinuously with periods of uplift and intervening periods of relative stability. In this way, the land rose slowly upwards with the effect that sea levels receded. Such rebound was not at the same rate everywhere; those parts of Britain that had been covered with the greatest thicknesses of ice, like the Highlands of Scotland, had been compressed the most and so uplifted to the greatest extent. Elsewhere, where the ice cover had been thinner, as for example south of the Severn-Thames line, isostatic uplift was to a much lesser degree. In this way, the land areas inundated during the Flandrian became progressively exposed, rising slowly from beneath the sea. The uplift was episodic with intervening periods of stability, since the sea was, in many places around the coast, able to cut and carve cliffs into the uplifted land. In this way, series of raised shorelines can be seen in many parts of Scotland, giving a step and stair, tread and riser-like form to the coastline, the oldest shoreline at the top and the youngest at the bottom, the last to emerge from beneath the sea (Figure 6.3). Another consequence of ice melting and isostatic recovery or uplift of the overlying crust is that the forebulge sinks back to its former level. This process contributed to the eventual demise of Cantref y Gwaelod and Lyonnesse and is the main reason why southern England, the most densely populated part of Britain, is still sinking slowly today; the sinking of the forebulge is not yet complete.

Figure 6.3 The former railway line to Inverbervie from Montrose was built for much of its length close to the coastal edge but along a raised shoreline. Now converted to a coastal path, this is, in places, under threat from wave erosion, as seen between Gourdon and Inverbervie, suggesting that relative sea level is now rising in this area (photo: R. W. Duck).

If, for example, the sea level is rising at a rate of 3 millimetres per year and the land level in a district is rising at an identical rate, then the net effect will be zero, a situation that is known as sea level stillstand. There will be no observed relative sea level rise or fall. Similarly, if sea level is falling and the land level is subsiding at an equal rate, stillstand conditions will prevail. Should the rate of uplift of the land exceed that of global sea level rise, coastal emergence will occur. However, in the opposite scenario of global sea level rise taking place at a greater rate than vertical uplift of the land, or indeed in the event of vertical land subsidence, the consequence will be coastal submergence and transgression of the sea over formerly dry land. If the lands adjacent to the coastline are relatively low-lying plains characterised by very gentle, low angle gradients, then even a small rise in relative sea level can potentially inundate large areas. The most vulnerable parts of Britain to relative sea level rise are thus the southern counties of England, areas where the ice cover was thin or absent and which are sinking due to forebulge compression, so the impacts of global sea level rise will be manifest more acutely. By contrast, in very broad terms, much of Scotland, where the thickness of Pleistocene ice was at its greatest, is still subject to isostatic uplift, which serves to moderate the effects of global sea level rise. This is, however, rather a simplification as there are large regional variations, even at the scale of an individual water body such as an estuary in which there can be differences in the rate of relative sea level change from the landward reaches to the mouth. Moreover, it is a popularly held rather broad-brush misconception that Scotland is 'bouncing back';[27] Caithness, the Orkney Islands, the Shetland Islands and the Outer Hebrides, are actually subject to relative sea level rise. In these north western and northern areas, the rate of global sea level rise exceeds isostatic uplift as there the ice cover during the Pleistocene was generally thin compared with elsewhere in Scotland.

The squeeze is on

Coastal environments are naturally dynamic and if we 'interpose a barrier against the raging ocean', for instance to form a railway, we – at least temporarily – constrain that dynamism and restrict the extent of tidal inundation or perhaps even the potential for landslides to occur. Where the relative sea level is rising, areas that are characterised by land claim embankments or sea walls and developments close to the shoreline are now experiencing what is called 'coastal squeeze'.[28] What would under natural conditions have been inter-tidal land has been enclosed from the influence of the sea and, where such structures are present, the position of high water mark is effectively fixed. As a result, any salt marsh that has been able to develop on the seaward side of the wall becomes squeezed against it and will, through time, disappear completely as there is no space for it to occupy.

The vegetation communities that develop on salt marshes often show a spatial zonation according to the elevation of the marsh surface, which controls both the time period and the frequency of inundation by tidal waters.[29] Different plant species can tolerate different periods and frequencies of submergence by salt water. This leads to a broad sub-division into what are often described as low marsh, mid marsh and high marsh, though the boundaries between these are gradational. Those plant communities that can tolerate the greatest degree of inundation grow on the low marsh, whereas those that can endure only very short or occasional, periodic submergence characterise the vegetation of the high marsh. On an area of natural coastline – one that is devoid of any physical barriers – these salt marsh zones and their associated plant communities can migrate inland through time in response to a rising relative sea level, which will result in a progressive change in the position of high water mark. In this way, the zone that was once occupied by high marsh plant communities might eventually become the more frequently and lengthily inundated low marsh, as the high marsh migrates and develops on top of once terrestrial environments.[30] The most up to date definition of coastal squeeze is given as:

> Coastal squeeze is one form of coastal habitat loss, where inter-tidal habitat is lost due to the high water mark being fixed by a defence or structure (i.e. the high water mark residing against a hard structure such as a sea wall) and the low water mark migrating landwards in response to sea level rise.[31]

Such areas are becoming progressively unable to accommodate erosion; inter-tidal sand or mudflats, slob and sleech, wetlands and salt marshes that would naturally migrate landwards as sea level rises are unable to do so as they encounter man-made barriers, hence the term 'squeeze'. Where a sea wall is present, although there might well be space for, say, low and mid marsh plant communities to migrate landwards, there is nowhere for the high marsh plants to go. So, the high marsh becomes squeezed against the sea wall and its plant communities ultimately disappear. If there is further sea level rise, the mid marsh plants may also become wiped out as the water depth and frequency of submergence becomes suitable only for the low marsh plants. In this way, lower marsh species might be obliterated and the former salt marsh becomes mudflat. The depth of water adjacent to the sea wall increases but so too the degree to which salt marsh vegetation is able to baffle wave energy becomes decreased. This permits waves of greater height to break against the structure, especially during storm conditions, increasing the likelihood of overtopping and potential flooding. Moreover, overtopping can often be the precursor to breaching of an embankment. If sediment availability to such areas has been depleted, as is so often the case by the construction of walls that cut off sediment input to the coast, then the squeeze is exacerbated. The result is that beaches become narrower and steeper in profile or indeed vanish completely due to erosion.

Our coastal railway embankments and walls are very important contributors to this process (Figure 6.4). The line to the resort

Figure 6.4 The vertical sea wall at Craigendoran on the north shore of the Firth of Clyde, now faced with concrete, carries a section of the railway line that connects Glasgow with Helensburgh. At high tide a patch of salt marsh is seen remaining squeezed against the sea wall. This structure is prone to overtopping by waves during severe storms, as in January 2014, and is one of many such walls, with railways on top, that are contributing to coastal squeeze (photo: R. W. Duck).

of Helensburgh from Glasgow opened in 1857 and was operated by the Edinburgh and Glasgow Railway Company.[32] Following the acquisition of the latter by the North British Railway Company in 1865, a proposal was made to build a pier for passenger steamers to land at Craigendoran in Dunbartonshire,[33] on the north bank of the Firth of Clyde on the outskirts of Helensburgh. Though there was opposition, the scheme went ahead and the pier was completed in 1881. This enterprise necessitated diverting the railway from its previous inland route at Craigendoran such that, to the south east of the new station, it was located on a stone wall right at the coastal edge. A violent storm hit the west coast of Scotland on 6 January 1882 and caused extensive damage to the Helensburgh line, most notably in the bay immediately to the south east of the railway station at Craigendoran. The brand new sea wall was no match for the force of the waves and was destroyed for a distance of between 200 and 300 yards. Furthermore, it was reported that, '. . . the permanent way has been almost wholly washed into the sea, while the rails were torn from the sleepers and lay at points mingled together in a confused heap.'[34] As a result of overtopping by waves, there was even considerable damage to the cement and brickwork of the new station building at Craigendoran.[35]

Elsewhere in western Scotland there was considerable damage to coastal railways, in particular at Saltcoats. The very name of this north Ayrshire town evokes its long association with the salt industry, fuelled by locally-won coal.[36] This industry began in the mid seventeenth century when the place 'Salt-cots, or Saltcoats, consisted of only four little cottages or cots, inhabited by as many families, who gained a livelihood by making salt in kettles.'[37] Opened in 1840, the railway eastwards towards the neighbouring town of Stevenston was built on a sea wall through the dunes along the back edge of the East Shore or East Beach, cutting off formerly inter-tidal lands from inundation. In the storm event of 6 January 1882, 30 feet of Saltcoats pier was said to have been washed away but, in addition:

The east shore experienced the full force of the storm. Previous storms had demolished a considerable part of the sea wall, and at several points threatened to damage the Glasgow and South-Western Railway. To guard against this workmen have been busy for weeks past strengthening the railway wall and constructing a breakwater with stones and timber at the most exposed parts.[38]

Their labours were, however, at least initially, not in vain.

The train from Glasgow due at 12.10 was looked for with great interest. At length, considerably behind time, the train was observed slowly approaching, the greatest caution being observed in going over that part of the line nearest the sea. The entire train was at times completely hidden from view by the dense showers of spray, but reached its destination in safety.[39]

Some half an hour after this train had passed, however, the sea broke through the wall in three places and washed over the line. At one point, 'no less than twenty yards of wall was washed down.'[40] Mid-way between Saltcoats and Stevenston, where there was extensive flooding in the town, another breach took place with the sea flooding a meadow – originally salt marsh – to landward of the line. Overtopping of the rails at both Craigendoran and Saltcoats was to become a frequent event[41] that often halted the passage of trains owing to the force of the waves.[42] When the sea wall at Craigendoran received the full force of a gale, it was often deemed inadvisable to permit trains to pass over it as it often sustained damage and the line was strewn with debris.[43] At the East Beach at Saltcoats, for instance, a storm that persisted for several days in early December 1893, 'tore up part of the breakwater near the Glasgow and South-Western Railway and swept away the footpath next the line. Great volumes of spray were blown across the line, but no injury was sustained by the permanent way.'[44] The following year, in December 1893, the footpath beside the railway at Saltcoats was again almost completely destroyed but the line itself remained intact.[45] In December 1900 another gale hit the East Beach and caused extensive damage to the sea wall protecting the line, carrying it away in several places. The sea was thus able to undermine the railway: 'For about 30 yards the rails of one of the lines were hanging with little support, the material underneath having been swept away.'[46]

Following several similar incidents,[47] a low, concrete promenade, about 800 metres in length was added in 1931, being built directly adjacent to the line at the East Beach.[48] The construction of its seaward edge, a near vertical wall, 'had not been without its engineering difficulties.'[49] However, as *The Scotsman* declared, 'It is now one of the finest promenades on the West Coast, and will be utilised by thousands of holidaymakers.'[50] Named 'Sandylands Esplanade', a reminder of the natural dunes that once characterised the area, it was thus to become

Figure 6.5 The 'Sandylands Esplanade' at Saltcoats is another contributor to coastal squeeze. Along with the adjacent Stevenston to Saltcoats railway line, it is frequently overtopped by storm waves (photo: R. W. Duck).

the first line of defence against the sea, exacerbating coastal squeeze but to a certain extent buffering the adjacent railway from storm waves (Figure 6.5). Nevertheless, overtopping of both the promenade and the railway at Saltcoats has proved a perennial problem.[51] Moreover, fast forward to December 2013 and both of these sections of railway sea wall continued to be savaged by the sea with overtopping by huge breaking waves taking place at Saltcoats and Craigendoran along with associated damage to overhead power lines.[52] To the east of Saltcoats, where the Sandylands Esplanade ends, the line diverges inland. Here, beyond the limit of coastal squeeze, the concrete structure has induced flanking erosion of the Stevenston beach sand dunes past its eastern extremity and the large bight created is being cut back inland so as to threaten the line (Figure 6.6). Incongruously, the very structure that affords a measure of protection along one section of the line might also be exacerbating its attack by the sea in another.

The coastal railway embankment that cuts off the mouth of the Ritec Estuary near Tenby is causing another effect, albeit temporary

Figure 6.6 Rubble dumped in an attempt to stem flanking erosion of the Stevenston beach dunes to the east of the slipway at the end of the 'Sandylands Esplanade', Saltcoats. The overhead power lines mark the proximity of the railway (photo: R. W. Duck).

in duration, which might be termed 'fluvial squeeze'. During periods of high river runoff – as in the floods that affected Pembrokeshire and beyond in January 2011, November and December 2012, and January 2014 – flood warnings were issued for the lower reaches of the River Ritec.[53] The railway impedes not only the natural tidal inundation by the sea but also the natural discharge to the sea and, in the event of high river discharges, holds back the flow of river waters causing flooding. Whilst the railway has helped to claim the estuary from the sea, the flooding of long claimed lands – 'The Marshes' – is today seen as another unwelcome consequence of the embankment.

In Wales alone, over 150 kilometres of the operating railway network is located close to the coastal edge, all of which is at risk of erosion or tidal flooding. A measure of the contribution that these railways are making to coastal squeeze is that about 100 kilometres of embankments and walls have the potential to prevent the landward migration of habitats in response to relative sea level rise.[54] The railway lines themselves – in Wales and elsewhere in similar coastal settings in Britain – are thus becoming more vulnerable to wave attack. And the dramatic storms of December 2013, through January 2014 to February 2014, irrespective of one's position on climate change, reaffirmed their vulnerability.[55] Whilst today we no longer experience the railway tragedies that took place on the edge of Britain in the mid to late nineteenth century, it is noteworthy that the breaches of embankments seen recently are by no means a novel phenomenon. Each of the sections of the lines that were severed by the sea during these gales has suffered a similar fate in the past, and repeatedly so since the lines were opened. The South Devon coast line through Dawlish, the South Wales coast line through Llanelli,[56] the Cambrian coast line to Pwllheli, the Cumbrian coast line from Carlisle to Barrow-in-Furness via Whitehaven; all had been ruptured by the sea repeatedly throughout their respective histories. Other important coastal arteries, such as the North Wales coastal route from Chester to Holyhead, survived the winter 2013–14 storms. This was, however, only by a hair's breadth as waves caused extensive damage to the promenade and the sea wall that protects the line at Deganwy near Llandudno.[57] Other parts of this line and indeed many coastal lines elsewhere in the country are also at risk of wave overtopping, erosion and breaching. Vulnerable sections not mentioned previously in the book include the Far North Line at Portgower, near Helmsdale in east Sutherland[58] and the final approach from Grimsby to the Lincolnshire terminus of Cleethorpes, built on land claimed from the sea. Whilst overtopped by waves and strewn with debris, the latter survived the 1953 storm surge but was severed the following year and by several subsequent events,[59] most notably another major surge that hit the east coast of England on 11 January 1978.[60]

Ironically, one railway is actually helping to create a new coastal wetland. Europe's largest construction project, London's new west-to-east artery known as Crossrail, is having an unusual impact at the coast. The scheme involves the twin boring of 21 kilometres of new tunnels beneath the capital.[61] Work began in September 2012 to move the uncontaminated spoils from these works to the Essex coast to create Europe's largest artificial nature reserve. The earth is being shipped to a jetty on Wallasea Island from which it is transferred to an 800-metre conveyor belt.[62] This island is today an almost perfectly flat piece of farm land at the confluence of two estuaries; that of River Crouch to the north and that of River Roach to the south. It is about 6.5 kilometres in length and 1.5 kilometres in width and is presently 2 metres below sea level at high tide. The land is surrounded by a sea wall that has claimed what were once five separate salt marsh islands for agriculture. This structure succumbed in several places to the 1953 storm surge, which caused almost complete inundation and devastation of the island. The floodwaters overtopped the wall and the island filled up with sea water like a huge basin. Waves breaking against the wall *from within* cut breaches from the vulnerable inner side, which, unlike its outer margin, was not protected by stone armouring.[63] The many ruptured sections of the embankment were subsequently repaired and the whole structure raised in level. More recently, however, a managed realignment scheme, to permit the sea to flood part of the island, thereby helping to accommodate sea level rise, and to regenerate salt marsh and mudflat has taken place.[64] The sea wall was breached in six locations in 2006 but a new wall was built further inland permitting tidal inundation over an area of a little over 100 hectares on the north eastern bank, thus easing the problems of flooding on the River Crouch. The spoils from the Crossrail excavations are being used to raise the level of almost the entire island by about 2 metres. A new landscape of meandering creeks, mudflats and salt marshes – last seen on the site 400 years ago prior to sea wall construction, drainage and land claim – will be created. This is simultaneously solving the two major problems of what to do with the vast quantities of earth, some 5,000,000–6,000,000 tonnes of clay, gravel and chalk, excavated from the tunnels beneath London and the huge cost of maintaining the flood defence wall that encircles Wallasea Island. The project, due for completion in 2020, is akin to building and landscaping a new island and the Royal Society for the Protection of Birds (RSPB), who own the reserve, aim to 'transform the island back into a magical intertidal coastal marshland.'[65] It is hoped that the 600 hectares of new habitats created through this unique project will support a striking range of nationally and internationally important bird populations, as well as a host of other wildlife. Railway construction is certainly still having previously unforeseen – but on this occasion positive – consequences on the vulnerable low-lying edge of Essex.

What does the future hold?

Whilst, as mentioned above, there will be considerable variations in the amount of relative sea level rise along the length and breadth of Britain, the recent predictions for the three capital cities, to 2095, are unequivocal. These suggest very similar, virtually identical rates of relative sea level rise for London and Cardiff according to high, medium and low greenhouse gas emissions scenarios. Relative to the 1990 sea level as a base, rises of 53.1, 44.4 or 37.3 centimetres are estimated for these two cities according to high, medium and low emissions scenarios, respectively.[66] For Edinburgh, owing to the greater amount of isostatic uplift, the rate of relative sea level rise according to the three scenarios through the twenty-first century is markedly lower; 39.2, 30.5 or 23.4 centimetres to 2095, respectively. During the period from December 2013 until February 2014, Britain was affected very severely by an exceptional set of winter storms that caused serious damage to coastal properties – not only railways – along with widespread, persistent flooding. The tidal surge of 6 December 2013 was the most serious in over 60 years since 1953 and gave rise to the highest tide that the Thames Barrier had experienced since it became operational in 1982. The Met Office's position is that:

> As yet, there is no definitive answer on the possible contribution of climate change to the recent storminess, rainfall amounts and the consequent flooding. This is in part due to the highly variable nature of UK weather and climate.[67]

Increased storminess and more unpredictable weather could lead to more frequent extreme events such as tidal surges. Severe storms around Britain have indeed become more frequent in the past few decades, although not above that seen in the 1920s.[68] So, to some extent the jury is still out with regard to the issue of increased storminess and the Met Office conclude that more robust research is required.[69]

Unplanned and unintended as it may be, railway formations – both operational and long defunct – in many parts of Britain now find themselves at the front line of coastal flood defence. A further irony is that, at a time when climate change is very much favouring rail as a means of both passenger and freight transport,[70] many of our already vulnerable coastal lines are becoming increasingly exposed to extreme weather and the very actions associated with their construction have exacerbated their susceptibility to attack by the sea. The breaches and disruptions to Britain's coastal railways seen in the first two months of 2014 will, for certain, happen again in the foreseeable future but not necessarily in precisely the same places or even on the same lines (Figure 6.7). The battle against the sea will never be won; some difficult decisions will have

Figure 6.7 This short section of the main East Coast railway line at Balmossie, between Broughty Ferry and Monifieth, is heavily defended by rock armour. It is one of the most vulnerable rail sections along the entire route given its proximity to the coastal edge. Wooden sea defences seen protecting the dunes in the background (looking east towards Monifieth) were uprooted, pushed over by breaking waves and damaged during the gales of December 2013 but the line was unaffected (photo: R. W. Duck).

to be made. Furthermore, just as every chain has its weakest link, every coast line will have its most vulnerable locations according to a particular set of weather conditions. More extreme weather conditions will also likely lead to increased frequency of coastal landslides and could, almost certainly, be a further catalyst for moving lines further inland. This could especially be so for strategically important routes such as that of the Cumbrian coast, which serves the spent nuclear fuel processing facility at Sellafield. It is now a very real possibility that the South Devon line between Exeter and Plymouth might be re-routed inland to avoid Dawlish. Several alternatives are being considered, one involving the utilisation of an old track bed to the north of Dartmoor that closed to traffic in the 1960s.[71] This diversionary scheme, along with the enormously controversial plans for High Speed 2 (HS2), is reawakening opposition to railways and will likely bring a new, twenty-first century wave of 'nimbyism' and land disputes to the table.[72] Nearly 170 years on, the spectre of several Dickensian earthquakes that could wholly change the law and custom of the neighbourhood has been rekindled.

Notes

1. Dickens, C. (1844), *The Life and Adventures of Martin Chuzzlewit*, London: Chapman and Hall.
2. 'Southborne-on-Sea' [sic], *The Hampshire Advertiser*, 3 October 1885.
3. Bainbridge, C. (1986), *Pavilions on the Sea: A History of the Seaside Pleasure Pier*, London: Robert Hale, 221 pp.
4. 'The severe gale', *The Standard*, 29 December 1900.
5. Bainbridge, C. (1986), *Pavilions on the Sea*.
    National Piers Society, History of Southbourne Pier, available at http://www.piers.org.uk/pierpages/NPSsouthborne.html (last accessed 26 April 2014).
6. Rennie, J. (1845), *Minutes of the Proceedings of the Institution of Civil Engineers*, 4, 24.
7. BBC News England (15 March 2014), 'Fire crews in Dawlish controlled landslip operation', available at http://www.bbc.co.uk/news/uk-england-26593362 (last accessed 26 April 2014).
    'Rebuilding Dawlish, swiftly and Lego-style', *The Sunday Times*, 16 March 2014.
    BBC News Devon (4 April 2014), 'Dawlish's storm-damaged railway line reopens', available at http://www.bbc.co.uk/news/uk-england-devon-26874503 (last accessed 26 April 2014).
8. Cole, M. (1997), 'Wave power', *New Civil Engineer*, 1242, 60–63.
9. Macfarlane, R. (2007), *The Wild Places*, London: Granta Books, 340 pp.
10. Price, M. R. C. (1989), *The Saundersfoot Railway*, Oxford: The Oakwood Press, 64 pp.
11. Price, M. R. C. (1989), *The Saundersfoot Railway*.
12. Price, M. R. C. (1989), *The Saundersfoot Railway*.
13. 'Disastrous gales and floods', *The Morning Post*, 9 October 1896.
    'The storm', *Glasgow Herald*, 9 October 1896.
14. 'Welsh coast erosion: Seaside village to be abandoned', *The Times*, 2 April 1936.
15. 'Welsh coast erosion: Seaside village to be abandoned', *The Times*, 2 April 1936.
16. Calculation made using: 'This is Money: Historic inflation calculator: how the value of money has changed since 1900', available at http://www.thisismoney.co.uk/money/bills/article-1633409/Historic-inflation-calculator-value-money-changed-1900.html (last accessed 26 April 2014).
17. 'Village defences smashed by sea', *The Scotsman*, 1 August 1938.
18. 'Heavy seas cause flooding', *The Times*, 25 September 1957.
    'Further damage by gales and floods', *The Times*, 26 September 1957.
19. 'Gales and high tides cause floods in Britain', *The Times*, 12 January 1974.
20. Williams, A. T., Ergin, A., Micallef, A. and Phillips, M. R. (2005), 'Public perception of coastal structures at groyned beaches', *Zeitschrift für Geomorphologie*, 141, 111–122.
21. Fairbourne Steam Railway, available at http://www.fairbournerailway.com/ (last accessed 26 April 2014).
22. BBC News Wales (11 February 2014), 'Sea level threat to force retreat of communities in Wales', available at http://www.bbc.co.uk/news/uk-wales-26125479 (last accessed 26 April 2014).
23. BBC News North West Wales (8 March 2014), 'Legal action threat over coastal

retreat of Fairbourne', available at http://www.bbc.co.uk/news/uk-wales-north-west-wales-26491477 (last accessed 26 April 2014).

24. Intergovernmental Panel on Climate Change (IPCC), available at http://www.ipcc.ch/ (last accessed 26 April 2014).

25. Climate Change: Evidence from the Geological Record. A Statement from the Geological Society of London (November 2010), available at http://www.geolsoc.org.uk/~/media/shared/documents/policy/Climate%20Change%20Statement%20final%20-%20new%20format.ashx (last accessed 26 April 2014).

    An Addendum to the Statement on Climate Change: Evidence from the Geological Record (December 2013), available at http://www.geolsoc.org.uk/~/media/shared/documents/policy/Climate%20Change%20Statement%20Addendum%202013%20Final.ashx (last accessed 26 April 2014).

26. *Climate Change 2013: The Physical Science Basis*, Chapter 13 – Sea Level Change, Intergovernmental Panel on Climate Change (IPCC), available at http://www.climatechange2013.org/images/report/WG1AR5_Chapter13_FINAL.pdf (last accessed 26 April 2014).

27. 'How Scotland is bouncing back: after 6,000 years, Ice Age phenomenon is keeping rising tides at bay', *Daily Mail*, 12 April 2000.

28. Doody, J. P. (2004), '"Coastal squeeze" – an historical perspective', *Journal of Coastal Conservation*, 10, 129–138.

    Doody, J. P. (2013), 'Coastal squeeze and managed realignment in south-east England, does it tell us anything about the future?', *Ocean and Coastal Management*, 79, 34–41.

29. Adam, P. (1990), *Saltmarsh Ecology*, Cambridge: Cambridge University Press, 465 pp.

30. French, P. W. (2001), *Coastal Defences: Processes, Problems and Solutions*, Abingdon, Oxfordshire: Routledge, 366 pp.

31. Pontee, N. (2013), 'Defining coastal squeeze: A discussion', *Ocean and Coastal Management*, 84, 204–207.

32. *Macneur & Bryden's (Late W. Battrum's) Guide and Directory to Helensburgh and Neighbourhood*, 7th Edition (1875), Helensburgh: Macneur & Bryden, 141 pp.

33. 'Proposed railway pier at Craigendoran, Helensburgh', *Glasgow Herald*, 29 January 1879.

34. 'Great Storm', *Glasgow Herald*, 7 January 1882.

35. 'Great Storm', *Glasgow Herald*, 7 January 1882.

36. Whatley, C. A. (1987), *The Scottish Salt Industry 1570–1850*, Aberdeen: Aberdeen University Press, 169 pp.

37. Chambers, R. and Chambers, W. (1836), *The Gazetteer of Scotland*, Edinburgh, 1031 pp.

38. 'Great Storm', *Glasgow Herald*, 7 January 1882.

39. 'Great Storm', *Glasgow Herald*, 7 January 1882.

40. 'Great Storm', *Glasgow Herald*, 7 January 1882.

41. 'Renewed gales', *Glasgow Herald*, 13 February 1884.

    'The gale in Scotland', *The Huddersfield Chronicle and West Yorkshire Advertiser*, 22 December 1900.

42. 'The severe gale', *The Morning Post*, 15 October 1891.

43. 'The severe gale', *The Morning Post*, 15 October 1891.

    'The Storm, Ten lives lost in Scotland', *Glasgow Herald*, 22 December 1900.

44. 'Severe storm', *Glasgow Herald*, 9 December 1893.
45. 'The storm', *Glasgow Herald*, 24 December 1894.
46. 'The storm', *Glasgow Herald*, 22 December 1900.
47. 'Severe gale: high tides and flooding', *The Scotsman*, 18 February 1910.
48. 'New esplanade: opened at Saltcoats. Sir Josiah Stamp's speech', *The Scotsman*, 27 July 1931.
49. 'New esplanade: opened at Saltcoats. Sir Josiah Stamp's speech', *The Scotsman*, 27 July 1931.
50. 'New esplanade: opened at Saltcoats. Sir Josiah Stamp's speech', *The Scotsman*, 27 July 1931.
51. 'Saltcoats. Streets flooded by sea', *The Scotsman*, 18 January 1934.
    Johnson, E. A., Guthrie, G., Battison, A., Sarker, M. A. and Hopewell, R. (2010), 'Saltcoats flood prevention scheme: Analysis, design and construction', *Coasts, Marine Structures and Breakwaters: Adapting to Change – Proceedings of the 9th International Conference, Edinburgh, 16–19 September, 2009*, 2, 120–131.
    Tozer, N., Pullen, T., Saulter, A. and Kendall, H. (2014), 'The Coastal Wave and Overtopping Forecast Service for Network Rail Scotland', in Allsop, W. (ed.) *From Sea to Shore – Meeting the Challenges of the Sea (Coasts, Marine Structures and Breakwaters 2013)*, London: ICE Publishing, in press, preprint available at: http://www.ice.org.uk/ICE_Web_Portal/media/Events/Breakwaters%202013/The-Coastal-Wave-and-Overtopping-Forecast-Service-for-Network-Rail-Scotland.pdf (last accessed 26 April 2014).
52. YouTube (24 December 2013), *Giant waves in Saltcoats, Scotland 2013*, available at http://www.youtube.com/watch?v=8j7qRAMe4NA (last accessed 26 April 2014). 'Storms batter Scotland', *Herald Scotland* (28 December 2013), available http://www.heraldscotland.com/news/home-news/storms-batter-scotland.23052049 (last accessed 26 April 2014).
53. BBC News Wales (14 January 2011), 'Disruption on road and rail in Wales as floods continue', available at http://www.bbc.co.uk/news/uk-wales-south-east-wales-12189145 (last accessed 26 April 2014).
    BBC News Wales (26 November 2012), 'Wales floods: Roads closed and trains cancelled', available at http://www.bbc.co.uk/news/uk-wales-20491996 (last accessed 26 April 2014).
    BBC News Wales (22 December 2012), 'Flooding: Heavy rain could spark more floods, agency warns', available at http://www.bbc.co.uk/news/uk-wales-20828329 (last accessed 26 April 2014).
    BBC News Wales (18 January 2014), 'Wales weather: Heavy rain as storm repairs begin', available at http://www.bbc.co.uk/news/uk-wales-25788710 (last accessed 26 April 2014).
54. Wales Coastal Monitoring Centre (2011), *First Annual Report 2010/11*, 66 pp., available at http://www.severnestuary.net/secg/docs/WCMC%20First%20Annual%20Report%202011.pdf (last accessed 26 April 2014).
55. BBC News, 'UK gales and storm surge – As it happened', available at http://www.bbc.co.uk/news/uk-25227714 (last accessed 26 April 2014).
56. *Llanelli Star* (9 January 2014), 'Damaged railway line "could reopen next week"', available at http://www.llanellistar.co.uk/Damaged-railway-line-reopen-week/story-20423277-detail/story.html (last accessed 26 April 2014).
57. *Daily Post* (16 January 2014), 'North Wales storms: Railway repairs urgently

needed say politicians', available at http://www.dailypost.co.uk/incoming/north-wales-storms-railway-repairs-6519642 (last accessed 26 April 2014).
BBC News Wales (6 February 2014), 'Flood risk as more rain lashes parts of Wales', available at http://www.bbc.co.uk/news/uk-wales-26055692 (last accessed 26 April 2014).

58. *John O'Groat Journal and Caithness Courier* (6 March 2013), 'Work starts on Portgower coast defence', available at http://www.johnogroat-journal.co.uk/News/Work-starts-on-Portgower-coast-defence-05032013.htm (last accessed 26 April 2014).

59. 'High tide stops town's rail services', *The Times*, 6 March 1954.
'Call for flood protection: Cleethorpes petition', *The Times*, 8 March 1954.
'High tides flood homes and cut railway lines', *The Times*, 5 January 1976.

60. 'Coast under threat, 1: How flooding and erosion take their toll. Counting the cost of keeping the sea at bay', *The Times*, 28 August 1978.
Steers, J. A., Stoddart, D. R., Bayliss-Smith, T. P., Spencer, T. and Durbidge, P. M. (1979), 'The storm surge of 11 January 1978 on the east coast of England', *The Geographical Journal*, 145, 192–205.
'Furious storms punch through resort's defences', *Grimsby Telegraph*, 18 January 2010, available at http://www.grimsbytelegraph.co.uk/Furious-storms-punch-resort-s-defences/story-11527454-detail/story.html (last accessed 26 April 2014).

61. BBC News: Science and Environment (17 September 2012) 'Wallasea Island nature reserve project construction begins', available at http://www.bbc.co.uk/news/science-environment-19598532 (last accessed 26 April 2014).
'Crossrail. Monster lift sends East London tunnelling machines 40 metres underground', available at http://www.crossrail.co.uk/news/articles/monster-lift-sends-east-london-tunnelling-machines-40-metres-underground (last accessed 26 April 2014).
Crossrail, 'Construction of Europe's largest man-made coastal reserve starts', available at http://www.crossrail.co.uk/news/articles/construction-europes-largest-man-made-coastal-reserve-starts (last accessed 26 April 2014).

62. Marinet, 'New London railway will result in new coastal wetland in Essex', available at http://www.marinet.org.uk/new-london-railway-will-result-in-new-coastal-wetland-in-essex.html (last accessed 26 April 2014).

63. Murphy, P. (2009), *The English Coast: A History and a Prospect*, London: Continuum Books, 282 pp.

64. Dixon, M., Morris, R. K. A., Scott, C. R., Birchenough, and Colclough, S. (2008), 'Managed realignment – lessons from Wallasea, UK', *Proceedings of the Institution of Civil Engineers Maritime Engineering*, 161, 61–71.
Macphail, R. I., Allen, M. J., Crowther, J., Cruise, G. M. and Whittaker, J. E. (2010), 'Marine inundation: Effects on archaeological features, materials, sediments and soils', *Quaternary International*, 214, 44–55.
Kadiri, M., Spencer, K. L., Heppell, C. M. and Fletcher, P. (2011), 'Sediment characteristics of a restored saltmarsh and mudflat in a managed realignment scheme in Southeast England', *Hydrobiologia*, 672, 79–89.

65. Royal Society for the Protection of Birds (RSPB), Wallasea Island Wild Coast project, available at http://www.rspb.org.uk/ourwork/casework/details.aspx?id=tcm:9-235089 (last accessed 26 April 2014).

66. UK Climate Projections: UKCP09, *Sea Level Rise*, available at http://ukclimateprojections.metoffice.gov.uk/21729 (last accessed 26 April 2014).
67. Met Office (February 2014), *A Global Perspective on the Recent Storms and Floods in the UK*, available at http://www.metoffice.gov.uk/research/news/2014/uk-storms-and-floods (last accessed 26 April 2014).
68. UK Climate Projections: UKCP09, *Observed Trends Report 1.6*, available at http://ukclimateprojections.metoffice.gov.uk/22864 (last accessed 26 April 2014).
69. Met Office and Centre for Ecology & Hydrology (February 2014), *The Recent Storms and Floods in the UK*, available at http://www.metoffice.gov.uk/media/pdf/n/i/Recent_Storms_Briefing_Final_07023.pdf (last accessed 26 April 2014).
70. Armstrong, J. and Preston, J. (2010), 'Rail in the context of climate change: strengths, weaknesses, opportunities and threats', *Proceedings of the 12th World Conference on Transportation Research, 11–15 July, 2010, Lisbon*, 15 pp.
71. BBC News England (12 March 2014), 'Storm-hit Dawlish: Where could a second rail line run?', available at http://www.bbc.co.uk/news/uk-england-26521168 (last accessed 26 April 2014).
72. 'High Speed 2 hell', *Evening Standard*, 16 October 2013.

Index

Note: **bold** page numbers refer to illustrations